Space Command Sustainment Review

Improving the Balance Between Current and Future Capabilities

Robert S. Tripp

Kristin F. Lynch

Shawn Harrison

John G. Drew

Charles Robert Roll, Jr.

Prepared for the United States Air Force

Approved for public release; distribution unlimited

RAND PROJECT AIR FORCE

The research described in this report was sponsored by the United States Air Force under Contract F49642-01-C-0003 and FA7014-06-C-0001. Further information may be obtained from the Strategic Planning Division, Directorate of Plans, Hq USAF.

Library of Congress Cataloging-in-Publication Data

Space command sustainment review : improving the balance between current and future capabilities / Robert S. Tripp ... [et al.].
       p. cm.
   Includes bibliographical references.
   ISBN 978-0-8330-4014-5 (pbk. : alk. paper)
   1. Astronautics, Military—United States. 2. United States. Air Force Space Command. 3. United States. Air Force—Equipment—Maintenance and repair. 4. United States. Air Force—Operational readiness. I. Tripp, Robert S., 1944–

UG1523.S633 2007
358'.8—dc22

2007009899

The RAND Corporation is a nonprofit research organization providing objective analysis and effective solutions that address the challenges facing the public and private sectors around the world. RAND's publications do not necessarily reflect the opinions of its research clients and sponsors.

RAND® is a registered trademark.

Cover design by Stephen Bloodsworth

Published 2007 by the RAND Corporation
1776 Main Street, P.O. Box 2138, Santa Monica, CA 90407-2138
1200 South Hayes Street, Arlington, VA 22202-5050
4570 Fifth Avenue, Suite 600, Pittsburgh, PA 15213-2665
RAND URL: http://www.rand.org/
To order RAND documents or to obtain additional information, contact
Distribution Services: Telephone: (310) 451-7002;
Fax: (310) 451-6915; Email: order@rand.org

Preface

This monograph examines options for improving Air Force Space Command (AFSPC) support and sustainment of U.S. Air Force space systems. Sustaining space capabilities is a complex undertaking involving preserving and protecting space launch capabilities, space vehicles, ground stations, and communications systems. It also encompasses the integration and augmentation of military capabilities with commercial and other government agencies' capabilities.

This monograph discusses the application of the strategies-to-tasks framework, a decision-support framework, to AFSPC maintenance and sustainment. We use an expanded strategies-to-tasks framework to explicate a maintenance and sustainment philosophy. Applying the strategies-to-tasks framework and this philosophy, we evaluate current space sustainment policies related to process, force development, doctrine, information systems and tools, and organization from a command perspective. From the same command perspective, we identify shortfalls and suggest, describe, and evaluate options for implementing improvements in current practices. Since space systems are diverse and since the analysis was limited to six months, we chose to use two example systems—the Global Positioning System and the Eastern and Western Range capabilities—to illustrate how the strategies-to-tasks framework can be applied across AFSPC sustainment practices.

AFSPC's Director of Air and Space Operations (AFSPC/A3) and Director of Logistics and Communications (AFSPC/A4A6) sponsored this research, which was conducted in the Resource Management Program of RAND Project AIR FORCE as part of a project entitled "Air

Force Space Command Logistics Review." The research for this monograph began in July 2005 and was completed in December 2005.

This monograph should be of interest to space logisticians, planners, acquisition personnel, and operators throughout the Department of Defense (DoD), especially those in the Air Force.

This monograph is one of a series of RAND documents that address agile combat support issues in implementing the air and space expeditionary force (AEF). Other publications issued as part of the larger project include the following:

- *Supporting Expeditionary Aerospace Forces: An Integrated Strategic Agile Combat Support Planning Framework*, by Robert S. Tripp, Lionel A. Galway, Paul Killingsworth, Eric Peltz, Timothy Ramey, and John G. Drew (MR-1056-AF), describes an integrated combat support-planning framework that may be used to evaluate support options on a continuing basis, particularly as technology, force structure, and threats change.
- *Supporting Expeditionary Aerospace Forces: New Agile Combat Support Postures*, by Lionel Galway, Robert S. Tripp, Timothy L. Ramey, and John Drew (MR-1075-AF), describes how alternative resourcing of forward operating locations can support employment timelines for future AEF operations. It finds that rapid employment for combat requires some prepositioning of resources at forward operating locations.
- *Supporting Expeditionary Aerospace Forces: A Concept for Evolving to the Agile Combat Support/Mobility System of the Future*, by Robert S. Tripp, Lionel Galway, Timothy L. Ramey, Mahyar Amouzegar, and Eric Peltz (MR-1179-AF), describes the vision for the Agile Combat Support (ACS) system of the future based on individual commodity study results.
- *Supporting Expeditionary Aerospace Forces: Lessons from the Air War Over Serbia*, by Amatzia Feinberg et al. (MR-1263-AF, not available to the general public) describes the Air Force's ad hoc implementation of many elements of an expeditionary ACS structure to support the air war over Serbia. Operations in Serbia offered opportunities to assess how well these elements actually supported

combat operations and what the results imply for the configuration of the Air Force ACS structure. The findings support the efficacy of the emerging expeditionary ACS structural framework and the associated but still-evolving Air Force support strategies.

- *A Combat Support Command and Control Architecture for Supporting the Expeditionary Aerospace Force*, by James Leftwich, Amanda Geller, David Johansen, Tom LaTourrette, C. R. Roll, Robert S. Tripp, and Cauley Von Hoffman (MR-1536-AF), outlines the framework for evaluating options for combat support execution planning and control (CSC2). The analysis describes the CSC2 operational architecture as it is now and as it should be in the future. It also describes the changes that must take place to achieve that future state.

- *Reconfiguring Footprint to Speed Expeditionary Aerospace Forces Deployment*, by Lionel A. Galway, Mahyar Amouzegar, and Don Snyder (MR-1625-AF), develops an analysis framework—as a footprint configuration—to assist in devising and evaluating strategies for footprint reduction. The authors attempt to define footprint and to establish a way to monitor its reduction.

- *Supporting Air and Space Expeditionary Forces: Lessons from Operation Enduring Freedom*, by Robert S. Tripp, Kristin F. Lynch, John G. Drew, and Edward W. Chan (MR-1819-AF), describes the expeditionary ACS experiences during the war in Afghanistan and compares these experiences with those associated with Joint Task Force–Noble Anvil, the air war over Serbia. This report analyzes how ACS concepts were implemented, compares current experiences to determine similarities and unique practices, and indicates how well the ACS framework performed during these contingency operations. From this analysis, the ACS framework may be updated to better support the AEF concept.

- *Supporting Air and Space Expeditionary Forces: A Methodology for Determining Air Force Deployment Requirements*, by Don Snyder and Patrick Mills (MG-176-AF), outlines a methodology for determining manpower and equipment deployment requirements for a capabilities-based planning posture. A prototype research tool, the Strategic Tool for the Analysis of Required Transportation, generates lists of capability units (unit type codes), which are required to support a user-specified operation.

- *Supporting Air and Space Expeditionary Forces: Lessons from Operation Iraqi Freedom*, by Kristin F. Lynch, John G. Drew, Robert S. Tripp, and C. R. Roll (MG-193-AF), describes the expeditionary ACS experiences during the war in Iraq and compares these experiences with those associated with Joint Task Force–Noble Anvil in Serbia and Operation Enduring Freedom in Afghanistan. This monograph analyzes how combat support performed and how ACS concepts were implemented in Iraq and compares current experiences to determine similarities and unique practices, and indicates how well the ACS framework performed during these contingency operations.

- *The Air Force Chief of Staff Logistics Review: Improving Wing-Level Logistics*, by Kristin F. Lynch, John G. Drew, David George, Robert S. Tripp, C. R. Roll, and James Leftwich (MG-190-AF), provides a review of Air Force base-level logistics processes. The review was designed to target process and process-enabler shortfalls that limited the ability of the logistics community to meet the increasing readiness demands. This monograph presents the background; the analytic approach, including the role RAND played; the results of that review; and the test and evaluation of solution options designed to improve wing-level logistics processes.

- *Supporting Air and Space Expeditionary Forces: Analysis of Combat Support Basing Options*, by Mahyar A. Amouzegar, Robert S. Tripp, Ron McGarvey, Edward Chan, and C. R. Roll (MG-261-AF), evaluates a set of global forward support location (FSL) basing and transportation options for storing war reserve materiel. The authors present an analytical framework that can be used to evaluate alternative FSL options. A central component of the authors' framework is an optimization model that allows a user to select the best mix of land- and sea-based FSLs for a given set of operational scenarios, thereby reducing costs while supporting a range of contingency operations.

- *Unmanned Aerial Vehicle (UAV) End-to-End Support Considerations*, by John G. Drew, Russell Shaver, Kristin F. Lynch,

Mahyar A. Amouzegar, and Don Snyder (MG-350-AF), presents the results of a review of current support postures for UAVs and evaluates methods for improving current postures that may also be applied to future systems.

- *Strategic Analysis of Air National Guard Combat Support and Reachback Functions*, by Robert S. Tripp, Kristin F. Lynch, Ronald G. McGarvey, Don Snyder, Raymond A. Pyles, William A. Williams, and Charles Robert Roll, Jr. (MG-375-AF), analyzes transformational options for better meeting combat support mission needs for the AEF. The role the Air National Guard may play in these transformational options is evaluated in terms of providing effective and efficient approaches in achieving the desired operational effects. Four Air Force mission areas are evaluated: continental United States centralized intermediate repair facilities, civil engineering deployment and sustainment capabilities, GUARDIAN[1] capabilities, and Air and Space Operations Center reachback missions.
- *A Framework for Enhancing Airlift Planning and Execution Capabilities Within the Joint Expeditionary Movement System*, by Robert S. Tripp, Kristin F. Lynch, Charles Robert Roll, Jr., John G. Drew, and Patrick Mills (MG-377-AF), examines options for improving the effectiveness and efficiency of intratheater airlift operations within the military joint end-to-end multimodal movement system. Using the strategies-to-tasks framework, this monograph identifies shortfalls and suggests, describes, and evaluates options for implementing improvements in current processes, doctrine, organizations, training, and systems.
- *Supporting Air and Space Expeditionary Forces: An Expanded Operational Architecture for Combat Support Planning and Execution Control*, by Patrick Mills, Ken Evers, Donna Kinlin, and Robert S. Tripp (MG-316-AF), 2006, expands and provides more detail on several organizational nodes in our earlier work that outlined concepts for an operational architecture for guiding the develop-

---

[1] GUARDIAN is an Air National Guard information system used to track and control execution of plans and operations, such as funding and performance data.

ment of Air Force CSC2 needed to enable rapid deployment and employment of AEF.

- *Combat Support Command and Control: An Assessment of Initial Implementations in Air Force Exercises*, by Kristin F. Lynch and William A. Williams (TR-356-AF), forthcoming, evaluates the progress the Air Force has made in implementing the TO-BE CSC2 operational architecture and identifies areas that need to be strengthened. Monitoring CSC2 processes, such as how combat support requirements for force package options needed to achieve desired operational effects were developed, assessment were made about organizational structure, systems and tools, and training and education.

RAND Project AIR FORCE

RAND Project AIR FORCE (PAF), a division of the RAND Corporation, is the U.S. Air Force's federally funded research and development center for studies and analyses. PAF provides the Air Force with independent analyses of policy alternatives affecting the development, employment, combat readiness, and support of current and future aerospace forces. Research is conducted in four programs: Aerospace Force Development; Manpower, Personnel, and Training; Resource Management; and Strategy and Doctrine.

Additional information about PAF is available on our Web site at http://www.rand.org/paf.

Contents

Figures

Tables

Summary

The ability to access and continuously operate in space is vital to economic, social, and military interests of the United States. Sustaining space capabilities is a complex undertaking. In this monograph, we examine options for improving AFSPC support and sustainment of U.S. Air Force space systems by evaluating the effectiveness and efficiency of current policies related to processes, force development, doctrine, information systems and tools, and organization from a command perspective.

The purpose of this monograph is to examine options for improving the sustainment of U.S. Air Force space systems, not by evaluating individual systems but by looking across AFSPC. By understanding current policies, we are able to suggest improvements in process, training and education, doctrine, systems and tools, and assignment of responsibilities from a command perspective. To this end, we used an expanded strategies-to-tasks framework as a "lens" for evaluating space system sustainment policies.[1] This expanded framework incorporates resource allocation processes and constraints in space system sustainment considerations. It also describes how space system sustainment resources and processes can be related to space capabilities and joint operational effects.

Finally, we evaluate options for improving space sustainment and provide both near- and longer-term implementation recommendations. Because space systems are very diverse and because the analysis time

---

[1] More-detailed information can be found in Appendix A.

frame was limited to six months, we use two example systems—the Global Positioning System and the Eastern and Western Range capabilities—to illustrate how the strategies-to-tasks framework can be applied across AFSPC sustainment practices.

Conclusions and Recommendations

The strategies-to-tasks framework provides a rationale for developing a commandwide philosophy for supporting space systems. The strategies-to-tasks framework prescribes separation of demand-side, supply-side, and integrator processes—which are often nested. Supply, demand, and integrator roles are not only defined at the execution level but also exist at other levels—both within and outside the command. Roles and responsibilities should be defined at all levels, stressing the importance of all three aspects of the strategies-to-tasks framework. Using a strategies-to-tasks framework and philosophy to separate supply, demand, and integrator processes to improve the effectiveness and efficiency of space system sustainment is an important first step. Once these responsibilities are separated, many other improvements can be made. The adoption of this philosophy can provide a basis for enhancing processes, force development, doctrine, information systems, and organization across the command that can be sustained over time and through many leadership changes.

Process Improvements

Once the strategies-to-tasks framework and philosophy have been adopted, many other processes improvements can be made. For example, the 30th Space Wing (SW) quality assurance (QA) process and supporting management information system (MIS) "best practices" could be adopted at other SWs. In addition, the QA MIS could be expanded to include system performance metrics, as well as the contract performance metrics it already contains. Reliability-centered maintenance practices should also be expanded within AFSPC beyond infrastructure-type equipment to include primary mission-equipment systems. Again, the 30th SW's experience in this area can provide the other wings with a model that they could tailor to their specific needs. (See pp. 21–37.)

The Integrated Priority List (IPL) used at the ranges could be expanded and implemented at nonrange wings. The process should be formalized to help wings identify, validate, approve, control, prioritize, and monitor the status of upgrades and modifications to space systems. In addition, a process for prioritizing and tracking sustainment resource requirements across the command (like a "super-IPL") could be established. This super-IPL could track all funding, including the sustainment money received from other agencies.

Finally, metrics need to be developed that relate sustainment resource needs to operational effects. Current metrics address individual system components and support with respect to that system. However, these metrics reflect past performance or effects or present report-card types of data. They do not provide predictive or leading indication of future issues that may arise within the space system. A focus on supply-and-demand metrics needs to be encouraged and can lead to improvements in metrics from the demand, supply, and integrator perspectives. The integrator may need an analytic arm to weigh demand- and supply-side inputs and provide a neutral viewpoint.

Force Development, Doctrine, and Information Systems and Tools Improvements

Development of civilian, officer, and enlisted logistics and communications leaders with space experience is essential to the success of the AFSPC mission. The strategies-to-tasks framework would suggest that maintenance be managed from supply-side organizations. This structure would provide a clear career path for support management and growth and provide a source for advocating career development and advancement up to commanding the maintenance group. (See pp. 39–44.)

A new program is being established with a developmental identifier to track space expertise as a step toward vectoring officers to AFSPC positions at appropriate points in their careers. Noncommissioned officer (NCO) development could benefit from more focused training in space systems to augment the Harris-type short courses, as well as some additional training and education in interpreting contractual documents for NCOs engaged in quality assurance evaluation. Space credentialing may also offer benefits for force management. At a minimum, special experience identifiers for space systems should be developed and applied to facilitate filling key space jobs in the officer and enlisted ranks. The civilian workforce could also benefit from credentialing space logisticians.

Although progress has been made in expanding doctrine to address support functions, support doctrine is not as robust as operational doctrine. In addition, space support doctrine is not as mature as aircraft support doctrine. AFSPC can make significant contributions by defining and inserting space support doctrine into Air Force publications and U.S. Strategic Command support doctrine into joint publications. AFSPC could work with Headquarters Air Force Directorate of Logistics Readiness, ACS Doctrine, and Wargames Division (AF/ILGX) to develop new Air Force Doctrine Document 2-4 subpublications and identify demand, supply, and integrator roles. The development of the strategies-to-tasks framework that outlines the philosophy for separating supply-side, demand-side, and integrator functions would contribute to doctrine and show how aircraft, missile, and space systems can follow the same philosophy in support, even if the specific implementations vary to some extent based on mission specifics, nature of systems, or history of support—organic or contractor—at both the service and joint levels.

Not all space systems provide the same level, in depth or breadth, of maintenance data collection and system and component status reporting as we observed in the Quarterly Sustainment Review or during similar forums for other systems. Standardization of data collection and reporting may be less important than standardized metrics. The critical issue is the ability to obtain key performance metrics so that trade-off decisions on where to invest sustainment dollars can be made if resources are constrained. These trade-off decisions should be based on collecting similar information about all the systems, so that like comparisons of metrics can be made.

More work can also be done on developing leading indicators. These indicators should underscore the cause of the resulting effect before the effect happens so that the indicators may be used for prediction. Decisionmakers should be able to focus their attention on future problems by using leading indicators.

In addition to metrics, sustainment actions and the effects of not performing them when necessary should be tracked. Many of the typical indicators (leading and lagging) can be used to prioritize sustainment actions better if properly integrated with acquisition schedules and operational schedules (planned). Others can be used to gauge contractor performance and award fees. Making trade-off decisions when resources are constrained requires having both measurements and metrics. Reliability-centered maintenance actions should be tied to these metrics as well.

Organizational Structure Improvements

Understanding the benefits of adopting the strategies-to-tasks framework and philosophy is important. If the benefits are explicitly recognized, organizational structure may not be as important. However, fully realizing the benefits of the philosophy based on the strategies-to-tasks framework means developing an organizational structure by following the philosophy. (See pp. 45–66 and Appendix B.)

Organizational changes at the headquarters have been made in accordance with the strategies-to-tasks framework, although this was not specifically recognized as strategies-to-tasks philosophy at the time. The organization of the space wings could follow suit—employing an integration function at either the group or wing, or by creating a maintenance group—to improve the balance between current readiness and future readiness to support operations.

Acknowledgments

Many people in the Air Force provided valuable assistance and support to our work. We thank Lt Gen Douglas Fraser (then Director of Air and Space Operations) and Brig Gen Thomas Deppe (then Director of Logistics and Communications, Chief Information Officer, and Chief Sustainment Officer), both from Headquarters AFSPC, for supporting this analysis. We would also like to thank Maj Gen William Shelton, 14AF Commander, and his staff for coordinating a review of this document in the five space wings that participated in this research (21 SW, 30 SW, 45 SW, 50 SW, and 460 SW). While we appreciate all their comments, the reviews concentrated on organizational options presented in the monograph and were very helpful in that aspect of the monograph.[1]

We are especially grateful for the assistance of Col Samuel Fancher, Louis Johnson, and Brian Healy. Colonel Fancher is Director of the Space Sustainment Division at Headquarters AFSPC. Mr. Healy is Deputy Director of the Space Sustainment Division. Mr. Johnson is Director of Logistics for the Space and Missile Systems Center. All three provided free and open access to their staff during our analysis and were invaluable in providing points of contact to aid in our research efforts.

This monograph would not have been possible without the support and openness of many individuals and organizations. We are especially grateful for the assistance given to us by Chris Milius, AFSPC/

---

[1]   14AF Staff Summary Sheet, RAND Space Command Maintenance Review: Balancing Current and Future Capabilities, March 2006.

A4SW, and Lt Col Richard Lawrence, AFSPC/A4SS. We also thank TSgt Thomas Oakes, AFSPC/A4SS; Mike Osborne, Command Logistics Manager for Ranges, AFSPC/A4S; and SMSgt Rodney Reyes, AFSPC/Space Professional Cadre. At 2 SOPS, we thank Capt Kevin Childs, Maj Theresa Malasavage, P. J. Mendicki, and MSgt Fred Woodring. At 50 SW, we thank Brian Wright, 850 SCS. At Vandenberg Air Force Base (AFB), California, we thank Lt Col Kevin Barker, 30 RMS Commander and Bruce Smith, Ronald Cockrell, John Hall, and Mike Huggins of the 30 RMS/RMO. We also thank James Walker, 30 RMS/RMQ, and Greg Packham, Call Henry, Inc. At Patrick AFB, Florida, we thank Lt Col Andrew Lester, 45 RMS Commander; Frank Davies, 45 RMS/RMOE; Gregg Kraver, 45 LCG/TD; TSgt Richard Allen, 45 LCG/LGQ; and TSgt Greg Jones, 1 ROPS. At SMC, we thank Col Richard Reaser, Jr., SMC/GP; Col Kenneth Robinson, SMC/GPG; Robin Pozniakoff, SMC/GPL; and Capt Timothy Barnes, SMC/GPL. At Buckley AFB, Colorado, we thank Col David Tobin, 460 OG Commander; Lt Col Daniel Wright III, 2 SWS; and Lt Phil Hoard, 2 SWS. At the 21 SW, we thank Col Donald McGee, Jr., 21 OG Commander; Col Lyman Faith, 21 MXG Commander; Lt Charles Waters, MXG/CCE; and Capt Joel Lane, 21 OG/CCE. We also thank Maj Robert Huckleberry and Brent Marley, Air University, Maxwell AFB, Alabama.

Our research has been a team effort with the Air Force Logistics Management Agency (AFLMA); this agency's support has been critical to this research. We thank Col Sean Cassidy, AFLMA Commander, and Lt Col Steven Purtle, AFLMA/LGM, for their support. We also especially wish to thank SMSgt Michael Dawson, AFLMA/LGM, for his assistance in collecting data and his thoughts and critiques of this analysis.

Finally, at the RAND Corporation, we thank our colleagues, David George, Patrick Mills, Don Snyder, and William A. (Skip) Williams for their contributions and critiques of our work. We would especially like to thank Myron Hura and David Ochmanek for their thorough review of this monograph. Their reviews helped shape this document into its final, improved form. Special thanks to Darlette Gayle for her support of this project.

Abbreviations

21M	missile maintenance career field
33S	communications maintenance career field
ACC	Air Combat Command
ACC/A3A	Air Combat Command, Directorate of Air and Space Operations, Airspace, Ranges, and Airfield Operations Division
ACC/A3T	Air Combat Command, Directorate of Air and Space Operations, Training Division
ACC/A6O	Air Combat Command, Communications Directorate, Communications Division
ACC/A7V	Air Combat Command, Installations and Mission Support Directorate, Environmental Division
ACS	Agile Combat Support
AEF	Air and Space Expeditionary Force
AFB	Air Force base
AFDD	Air Force doctrine document
AFI	Air Force instruction
AFMC	Air Force Materiel Command
AFPD	Air Force policy directive

AFRC	Air Force Reserve Component
AFSC	Air Force specialty code
AFSCN	Air Force Satellite Control Network
AFSPC	Air Force Space Command
AFSPC/A3	Air Force Space Command, Directorate of Air, Space, and Information Operations
AFSPC/A3C	Air Force Space Command, Space Superiority Operations Division
AFSPC/A3F	Air Force Space Command, Global Space Operations Division
AFSPC/A3N	Air Force Space Command, Nuclear and Helicopter Operations Division
AFSPC/A3R	Air Force Space Command, Assured Access Operations Division
AFSPC/A3T	Air Force Space Command, Training, Test, Exercises, and Evaluation Division
AFSPC/A4A6	Air Force Space Command, Directorate of Logistics and Communications
AFSPC/A4S	Air Force Space Command, Space Systems Sustainment Division [formerly AFSPC/LCZ]
AFSPC/A4SM	Air Force Space Command, MILSATCOM Division
AFSPC/A4SP	Air Force Space Command, Policy and Programs Division
AFSPC/A4SS	Air Force Space Command, Spacelift and Range Division
AFSPC/A4SW	Air Force Space Command, Warning and Surveillance Division

AFSPC/A5	Air Force Space Command, Directorate of Requirements [formerly AFSPC/DR]
AFSPC/A7	Air Force Space Command, Directorate of Mission Support [formerly AFSPC/MS]
AFSPC/A8A9	Air Force Space Command, Directorate of Plans, Programs, Analysis, Assessment and Lessons Learned [formerly AFSPC/XP]
AFSPC/DRS	Air Force Space Command, Space Support Division
ANG	Air National Guard
AOR	area of responsibility
AS-IS	The system in its form as of the time of this research
ASTS	Air and Space Test Squadron
C4ISR	command, control, communications, computers, intelligence, surveillance, and reconnaissance
CAMS	Core Automated Maintenance System
CBM+	condition-based maintenance plus
CC	commander
COCOM	combatant command
CSC2	combat support execution planning and control
CWO	combat wing organization
DISA	Defense Information Systems Agency
DO	operations officer
DoD	Department of Defense
DSP	Defense Support Program

ESC   Electronic Systems Center

FYDP   Future Years Defense Program

GETS   GPS Enhanced Theater Support

GPS   Global Positioning System

HF   helicopter flight

HQ   headquarters

ICBM   intercontinental ballistic missile

IMDS   Integrated Maintenance Data System

IPL   Integrated Priority List

ISR   intelligence, surveillance, and reconnaissance

JTF   joint task force

LCG   launch control group

LCSS   launch support squadron

MAJCOM   major command

MOF   maintenance operations flight

MSG   mission support group

MTBF   mean time between failures

MTTR   mean time to repair

MXG   maintenance group

NASA   National Aeronautics and Space Administration

NGA   National Geospatial-Intelligence Agency

NOG   network operations group

O&M   operations and maintenance

OG   operations group

OSS	operations support squadron
PAF	RAND Project AIR FORCE
PMD	program management division
POM	Program Objective Memorandum
PTR	Primary Training Range
QA	quality assurance
QAE	quality assurance evaluation
QSR	quarterly sustainment review
RCM	reliability-centered maintenance
REMIS	Reliability and Maintainability Information System
RMA	reliability, maintainability, and availability
RMO	range management office
RMS	range management squadron
RMS/RMQ	Range Management Squadron, Quality Assurance Flight
ROPS	range operations squadron
SCF	space communications flight
SCS	space communications squadron
SLS	space launch squadron
SMC	Space and Missile Systems Center
SMS	space management squadron
SMXS	space maintenance squadron
SMU	space maintenance unit

SOPS	space operations squadron
SPCS	space control squadron
SW	space wing
SWS	space warning squadron
TDS	Theater Distribution System
TO-BE	The system in its future target form
TTP	tactics, techniques, and procedures
UAV	unmanned aerial vehicle
UMD	unit manpower document
USecAF	Under Secretary of the Air Force
USSTRATCOM	U.S. Strategic Command
WRM	war reserve materiel
WS	weather squadron

CHAPTER ONE

Introduction

The ability to access and continuously operate in space is vital to the economic, social, and military interests of the United States. Sustaining space capabilities is a complex undertaking involving preserving and protecting space launch capabilities, space vehicles, ground stations, and communications systems, and the integration and augmentation of military capabilities with commercial and other government agencies' capabilities and with joint operational effects for the military. This monograph evaluates options for improving the Air Force Space Command's (AFSPC's) support and sustainment of U.S. Air Force space systems by assessing the effectiveness and efficiency of current policies related to processes, organization, force development, doctrine, and information systems and tools from a command perspective.

Study Motivation and Scope of the Analysis

In 2005, Maj Gen Douglas Fraser, then–Director of Air and Space Operations (AFSPC/A3), and Brig Gen Thomas Deppe, then–Director of Logistics and Communications (AFSPC/A4A6), observed that there did not appear to be a common, commandwide logistics philosophy for governing the sustainment of space systems. They asked RAND Project AIR FORCE (PAF) to evaluate AFSPC's logistics policies from an end-to-end perspective, to include space vehicles, ground stations, and communication networks connecting space platforms and command-and-control centers from a command perspective. They were interested in whether existing support philosophies and approaches could be improved to ensure that space capabilities are preserved and protected now and in the future. Across systems, some of the space-support planning and execution policies questioned included the following:

- consistency of maintenance philosophy and standardized policies
- ability to predict sustainment requirements and unfunded surprises
- ability to relate sustainment funding to operational effectiveness
- force development of support personnel
- oversight and insight of contractor support actions
- aging ground systems
- seams between sustainment and acquisition.

Generals Fraser and Deppe asked us to take a broad, objective examination of how to improve space-system support but to keep our analysis of options to those that would be feasible within expected funding levels. They also asked that we separate improvement options into those that could be implemented in the short term and those that would take longer to implement. The longer-term initiatives could require enabling technology that has not yet been developed or that would require significant time to modify systems, or other such reasons. PAF was asked to complete the review within six months and to consider all appropriate stakeholders in the analysis, including

- Headquarters (HQ) AFSPC
- AFSPC wings in the 14th Air Force (AF)
- the centers: the Space and Missile Systems Center (SMC) and the Electronic Systems Center (ESC)
- other users of space capabilities
- other operators of space systems, including commercial and other government entities.[1]

---

[1] We did not include this group of stakeholders, although we assume that we understand some of their concerns.

Missile maintenance and support were not considered as part of this study. We focused the analysis on space systems, which have different characteristics from those of aircraft and missile systems. For the most part, space systems are fielded in small numbers. For example, the Global Positioning System (GPS) requires 24 satellites in orbit, four ground antennas, six monitoring stations, and one master control station. In contrast, the F-16 has 1,150 aircraft in the active and reserve components.[2]

There is a large variance in the ages of space systems—very new to very old. Because space systems are so expensive, they are kept in service as long as possible. Some, such as the AN/FPS-85 phased-array radar system, are well over 40 years old. Some portion of the original missions of such older systems may be performed by other, newer space assets or, perhaps, is no longer performed at all. And although there is some redundancy among space systems, most systems are designed to meet unique needs. Therefore, some of the benefits of developing standardized procedures and policies (such as standardized maintenance procedures and training approaches) for aircraft and missile systems with large inventories of common operational platforms may not benefit these few or one-of-a-kind platforms and missions. In addition, most space-system sustainment maintenance is performed by contractors, who have separate or distinct capabilities, making standardization difficult.

The operational tempo for space systems is different from those for aircraft and missile systems. GPS course corrections may occur once a day; however, both military personnel and civilians rely on GPS capabilities 24 hours a day, seven days a week. The ranges may have 14 launches a year, but launch preparation lead times extend for months. There is more time between major events than there is for flying F-16 sorties (whether in training or, especially, in combat operations), but such capabilities as secure satellite communications are required almost continually.

Maintenance is also performed differently on space systems than on ground-based systems. Some space systems require on-orbit main-

---

[2] *Air Force Magazine: 2005 Air Force Almanac.*

tenance. They do not fly back to base for repair as aircraft do. Maintenance is performed remotely, while the system is still operating.

Space support organizations are different from other Air Force support organizations. HQ AFSPC has combined logistics and communications into AFSPC/A4A6 (formerly AFSPC LC). HQ U.S. Air Force has separate logistics and communications directorates. Space wings (SWs) have logistics embedded in the operations group (OG), whereas fighter wings have separate maintenance and operations groups.[3] One acquisition agency, SMC, has been embedded in AFSPC, while another, ESC, has not.

In addition to the Air Force, the many users of space-system capabilities include commercial vendors, other U.S. government agencies, and other governments. The Air Force supports some of these activities, including bearing some of the costs. Some commercial satellite launches use Air Force ranges. Other U.S. government agencies, such as the National Reconnaissance Office, use Air Force launch facilities and ground station capabilities to operate their systems.

Besides having unique characteristics, AFSPC space operations and sustainment directly affect military capabilities. AFSPC is charged with supporting the following mission areas:

- force enhancement—communications; weather; intelligence, surveillance, and reconnaissance; navigation; and missile warning, all providing support to combatant commands (COCOMs)
- space control—space situational awareness and counterspace
- force application—intercontinental ballistic missiles (ICBMs)
- space support—launch and satellite operations providing support to other U.S. government agencies and commercial vendors.

Each of these mission areas directly affects military capabilities. For example, consider how space assets support the development of the Theater Distribution System (TDS) to meet COCOMs' needs in

---

[3]   Although 21 SW has a maintenance group (MXG), logistics is spread throughout both the operations and maintenance areas.

their areas of responsibility (AORs).[4] All TDS planning and execution activities require access to space-based systems to carry communications between forward- and rearward-based elements of the COCOM staff. They also require access to the intelligence and weather information space assets provide.

Another example is the use of the MQ-1 Predator unmanned aerial vehicle (UAV) in military operations. The first variant, the RQ-1, was designed to provide persistent intelligence, surveillance, and reconnaissance (ISR) coverage of a specified target area. The current variant, the MQ-1, was designed to support time-sensitive targeting—continuously monitoring suspected target areas and promptly attacking targets when they emerge. Both models are designed to loiter over a target area and relay relatively high-resolution pictures and video of specific targets on the ground. The data the UAV collects over the target area, as well as air vehicle command signals, are relayed back through the ground control station using satellite links. Without satellite communications, UAVs would be limited to line-of-sight operations.[5]

Air Force space-system support actions not only affect military capabilities but also affect national goals. Some military systems, such as weather systems and GPS, provide services for public and government uses. The Air Force provides GPS and weather services as public goods and bears the full costs of sustainment for all users.

The combination of the uniqueness of space-system characteristics, the importance of space systems to military and commercial capabilities, and national goals presents challenges for space-system sustainment.

---

[4] TDS development includes developing an airlift network, deploying communications and information systems needed to manage and control airlift operations, deploying and sustaining resources needed to run air terminal operations, and providing the combat support resources required for housing deployed airlift operations at forward and main operating bases.

[5] *Line-of-sight* refers to a direct link from a ground control station to an airframe (Drew et al., 2005).

Organization of This Monograph

In Chapter Two, we discuss our analytic approach and the strategies-to-tasks framework. In Chapter Three, we apply the expanded framework to space-system sustainment to explicate a space sustainment philosophy and discuss shortfalls in the AS-IS space sustainment philosophies and processes. We also address actions that can be taken to address the shortfalls. Chapter Four discusses shortfalls in force development, doctrine, and information systems and tools. In Chapter Five, we discuss organizational options. Chapter Six contains our conclusions and a discussion on how to proceed with the suggested changes. Appendix A presents the basic strategies-to-tasks framework. Appendix B presents organizational structure options for HQ AFSPC and the 21, 30, and 460 SWs. Appendix C contains manpower data for the 30, 45, and 50 SWs. Appendix D summarizes range service-support contracts, while Appendix E compares them. Appendix F then provides an example of the reliability centered maintenance (RCM) prioritization processes. Appendix G describes 21 MXG. And finally, Appendix H describes the evolution of space wing maintenance organizations.

CHAPTER TWO

Analytic Approach and the Strategies-to-Tasks Framework

The analysis in this monograph focuses on AFSPC's sustainment of space systems. Sustainment processes include

1. maintaining such equipment as satellite buses, payloads, ground stations, communications systems, and other infrastructure necessary to provide space capabilities
2. managing materiel, providing spare parts where and when needed[1]
3. maintaining and sustaining the space launch ranges
4. contracting sustainment processes, providing commercial capabilities for maintenance, materiel management, and other services to support space sustainment.

The Air Force is responsible for sustaining unique systems for Air Force use and for sustaining some systems for other agencies, such as the National Aeronautics and Space Administration (NASA) and the National Reconnaissance Office. Some military systems, such as GPS, serve not only the military but also commercial and civic enterprises. In many cases, the Air Force does not control all the components of the end-to-end system.

The development of improvement options must recognize that Air Force space operations are part of an integrated, end-to-end, multi-

---

[1] Although materiel management is part of sustainment, this study did not address this area.

component system. The components of that system include space vehicles, ground stations, command-and-control facilities, launch facilities, and communications links among space vehicles, command-and-control hubs, and users. Efforts to improve sustainment of U.S. Air Force space systems should also take into account the roles of joint providers and incorporate them into the options.

Since space systems are diverse and since the research for this effort was limited to six months, we have used two example systems—GPS (a specific system) and Eastern and Western Range capabilities (a broader capability)—to illustrate sustainment practices across AFSPC. While we limited our analysis to these two areas, we could draw general conclusions about overall space-system sustainment from these investigations. We selected these systems for several reasons:

- First, GPS is a relatively well-defined system with a clear mission. SMC developed the system, and the 2nd Space Operations Squadron (2 SOPS) at 50 SW operates it.
- Second, the Eastern Range, at Patrick Air Force Base (AFB), Florida, and the Western Range, at Vandenberg AFB, California, have both launched GPS satellites. These ranges provide capabilities that are necessary for launching and testing many systems. Thus, they share some common characteristics with other space systems, such as radars at ground stations, and also have some unique features.
- Third, these two systems present a wide spectrum of characteristics and requirements that can be used to illustrate a command perspective of space-system sustainment. An analysis of these systems appears to be broad enough on which to base general observations and different enough to ensure that unique aspects of space systems are not overlooked when considering sustainment improvement options.

Analytic Approach

To conduct this review, we worked with key stakeholders and used a structured methodology to identify specific areas of analysis.[2] The first step was to gather data and inputs by visiting AFSPC units, including each of the five space wings in 14AF, SMC, the COCOMs, ESC, the Air Staff, other space users, and the acquisition communities. We worked with many individuals, both inside and outside the Air Force (see Table 2.1 for a list of the key stakeholders). Each organization discussed issues associated with space-system sustainment from its vantage point.

Following the framework outlined in Figure 2.1, the PAF analysis team evaluated data collected from key stakeholders, as well as written materials, and grouped the information into nine major categories:

1. maintenance philosophy
2. maintenance management
3. contracting management
4. logistics planning
5. material management
6. force development[3]
7. information systems and tools
8. sustainment problem prediction
9. metrics.[4]

Through further analysis and data collections, we eventually grouped the data into four major focus areas for more-detailed analysis. The four main focus areas examined enhancements to

---

[2] The Chief of Staff's Logistics Review, conducted in 2002, used a similar methodology. See Lynch et al. (2005) for more details.

[3] This category includes officer and enlisted training and career progression.

[4] The metrics include leading indicators of sustainment problems and the ability to relate sustainment needs to operationally relevant metrics for use in the Program Objective Memorandum (POM) process.

Table 2.1
Research Key Stakeholders

Organization	Location	Units
AFSPC/A3[a]	Peterson AFB, Colorado	
AFSPC/A4A6[b]	Peterson AFB, Colorado	
AFSPC/A4S[c]	Peterson AFB, Colorado	
14AF	Vandenberg AFB, California	
21 SW	Peterson AFB, Colorado	21 OG 21 MXG
30 SW	Patrick AFB, Florida	30 Range Management Squadron (RMS)
45 SW	Vandenberg AFB, California	45 Launch Group 45 RMS
50 SW	Schriever AFB, Colorado	2 SOPS 850 Space Communications Squadron (SCS)
460 SW	Buckley AFB, Colorado	460 OG 2 Space Warning Squadron (SWS)
SMC	Peterson AFB, Colorado Los Angeles AFB, California	Logistics Group (LG) GPS Joint Program Office
Air University	Maxwell AFB, Alabama	
Contractors	All locations	

[a] Formerly AFSPC/XO.
[b] Formerly AFSPC/LC.
[c] Formerly AFSPC/LCZ.

1. logistics philosophy, process, and organization
2. force development
3. information systems
4. prediction, metrics, and the POM planning process.

The focus areas provided a foundation for analyzing solution options with respect to their current and future impact on space-system sustainment.

Figure 2.1
Areas of Analysis and Specific Focus Areas

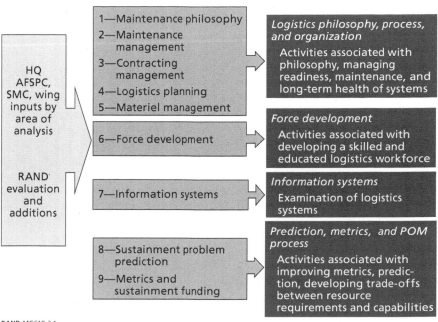

RAND *MG518-2.1*

The purpose of this monograph is to evaluate options for improving AFSPC sustainment of U.S. Air Force space systems from a command perspective, rather than by individual system. Understanding current policies has enabled us to suggest improvements in process, training and education, doctrine, systems and tools, and assignment of responsibilities. To this end, we used an expanded strategies-to-tasks framework as a "lens" for evaluating space-system sustainment policies.[5] This expanded framework incorporates resource allocation processes and constraints into space-system sustainment considerations. It also describes how space-system sustainment resources and processes can be related to space capabilities and joint operational effects.

We used this expanded strategies-to-tasks framework to examine current, or AS-IS, command philosophies and to explicate a space

---

[5] The strategies-to-tasks framework is discussed at the end of this chapter. More details can be found in Appendix A.

maintenance and sustainment philosophy. Applying this proposed philosophy, we evaluated policies related to processes, force development, doctrine, information systems and tools, and organizational structure. We identified disconnects and gaps in AS-IS policies and, in this monograph, suggest, describe, and evaluate options for implementing improvements in current practices (see Figure 2.2).

Finally, in keeping with the guidance received from our sponsors, we evaluated options for improving space sustainment and here provide both near- and longer-term implementation recommendations.

The Strategies-to-Tasks Framework

The strategies-to-tasks framework was developed at the RAND Corporation during the late 1980s (see Kent, 1989, and Thaler, 1993). It has been widely applied in the Department of Defense (DoD) to aid in strategy development, campaign analysis, and modernization planning.[6] The framework has proven to be a useful approach for providing intellectual structure to ill-defined or complex problems. Working through the strategies-to-tasks hierarchy can help identify areas where new capabilities are needed, clarify responsibilities among actors contributing to accomplishing a task or an objective, and place into a common framework the contributions of multiple entities and organizations working to achieve some common objective.

In this analysis, we used an expanded strategies-to-tasks framework to highlight the task of resource allocation. The basic resource allocation task for space-system sustainment activities can be viewed as a problem of integrating the *demand* for resources (that is, processes associated with using space-system capabilities) with the *supply* (that is, processes associated with planning, replanning, and executing

---

[6]   Internal examples are Lewis, Coggin, and Roll (1994) and Niblack, Szayna, and Bordeaux (1996). Outside RAND, versions of the framework are in use by the Air Force, the Army, and elements of the Joint Staff.

Figure 2.2
Space Command Sustainment Review Methodology

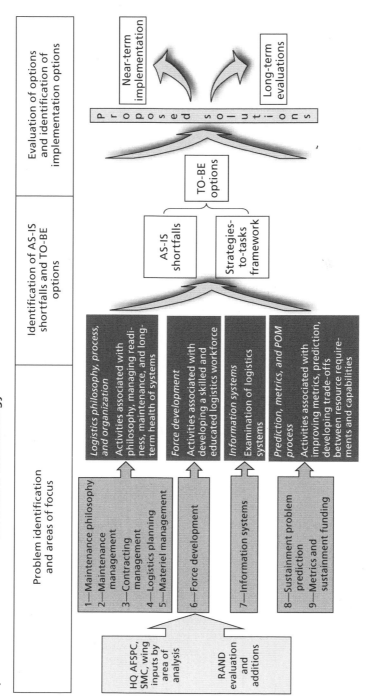

space-system operations to support a COCOM requirement in a specific AOR).[7] Finally, we identified *integrator* processes (those associated with allocating scarce space resources to prioritized COCOM needs for accomplishing space-support objectives) and will show how space-system sustainment can be related to task-organized operational elements used to create desired joint operational effects.

Figure 2.3 illustrates how resource allocation considerations can be integrated into a strategies-to-tasks framework that manages space-system sustainment processes. This framework is important for understanding the philosophy that should guide logistics and operations processes, force development, doctrine, information systems and tools, and organization.

Figure 2.3
Strategies-to-Tasks Framework with Resource Allocation Considerations

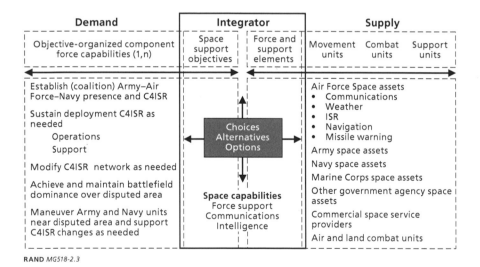

RAND *MG518-2.3*

---

[7]   Examples of *demand-side* processes include reviewing and prioritizing operational schedules in terms of COCOM and other governmental priorities, assessing the effectiveness of system performance, validating operational needs, assessing the timeliness of requirements submissions, and deconflicting competing operational needs. Examples of *supply-side* processes include determining sustainment needs, analyzing sustainment options, identifying sustainment schedules and any operational downtime, and monitoring sustainment performance.

The left side of Figure 2.3 lists some illustrative space-support demand-side objectives. The right side lists some illustrative force and support elements that can be selected from component providers to satisfy the objectives on the left. The middle shows the integration of demand and supply processes. Here, the integrator chooses the force and support elements from the available options, each of which may have different attributes and different operational effects. The result creates the space capabilities shown at the bottom of the integration box.

In this case, the choices can be made from a set of options that includes space assets the service provides or that are available from commercial and other government sources. Each choice may result in differing capabilities, for example, different timelines for establishing coalition or joint Army and Air Force presence in the area of interest.

The support of space objectives creates the demand to be satisfied, as determined by the integrator. Each supply element is generally provided by a component. Each space-support objective may require combinations of component resources to achieve the desired capability and, ultimately, the joint operational effect. For example, a COCOM may need assured, secure satellite communications in several parts of the AOR. There may be more than one way to meet that demand— component assets and/or commercial assets. The neutral integrator would decide how to best meet all or part of the demand for satellite communications using the all the space-system resources available.

To balance competing requirements (between the demand for space-system support and the available space-system resources), we considered two principles. First, supply-side and demand-side decision-making processes should be independent of one another, with the integrator making the choices. Goals and objectives should also be developed independently. We call this the *independence principle*.

If the integrator is too close to the supply side, decisions may be affected more by ensuring that sustainment actions are taken when they are requested rather than when they are required. Insufficient attention may be given to the operational needs of launching new space capabilities or using ground stations and command-and-control centers to perform current operations. If, on the other hand, the integrator is too close to the demand side, current operations may always be given first priority, and not enough time may be devoted to performing sustainment actions.

With respect to space capabilities and the very high dollar value of these systems, sustainment actions may need to take second priority. That is, space operations may always need to take precedence. For example, GPS Enhanced Theater Support (GETS) requests from COCOMs or Space-Based Infrared System (460 SW) responses to Integrated Tactical Warning and Attack Assessment directives may supersede scheduled sustainment activities needed on the satellite or at the ground station. If so, the supply side must know about future requirements and develop flexible processes to perform sustainment actions around the requirements. If the independence principle is violated, those in senior leadership positions need to be aware that the principle is being violated and need to plan how to compensate to mitigate the potential resulting effects.

The second principle suggests that supply-side and demand-side decisions should be made separately. Following this principle, the demand side specifies operational requirements and priorities for ground stations and command-and-control facilities (for example, launching of future systems, replacement of space-based assets, and modification of support and infrastructure). The supply side decides how to satisfy those needs. So, the demand side does not instruct the supply side when and how to schedule sustainment actions but rather informs it of when specific capabilities are needed, to the extent that they are known. The supply side determines the sustainment actions and schedule needed to satisfy the operational requirements within the time frame needed.

Applying the principles of separation and independence in the strategies-to-tasks framework creates a tension between the supply side and the demand side. This tension is natural and needs to be explicitly recognized by senior leaders. Once it has been recognized, processes and organizations can be established to leverage this tension.

As an illustration, consider the philosophy that was used to establish the combat wing structure for fighter wings. In the objective wing, before implementation of the combat wing structure, many supply-side functions (organizational maintenance) were embedded in demand-side organizations (the OG). Thus, the objective wing structure was not separate or independent. As senior commanders pointed out (Jumper, 1999; Gabreski, 2000), maintenance requirements leaned toward providing readiness *now*. Long-term maintenance requirements were not receiving enough attention.

In response to indicators of declining readiness, heightened operational tempo, and evolving force employment concepts, then–Chief of Staff of the Air Force Gen Michael E. Ryan chartered a review of wing-level logistic processes, called the Chief's Logistics Review (Lynch et al., 2005), which ultimately led to the formation of the combat wing structure under then–Chief of Staff Gen John P. Jumper.[8] The combat wing structure recognizes the strategies-to-tasks principles of independence and separation and places all supply-side maintenance activities under a maintenance group commander and all demand-side activities under the OG commander. The wing commander, who acts as the integrator, arbitrates the tension created between the two competing sets of requirements, supply side and demand side. In this structure, the wing commander makes choices about meeting operational and sustainment requirements.

With these principles in mind, Figure 2.4 shows a high-level view of a strategies-to-tasks organizational approach that shows various supply, demand, and integrator roles. This view provides insights on space-system planning and processes at the execution level and assigns processes among existing organizations.

In this execution-level representation, the Under Secretary of the Air Force acts as the integrator, providing integrated space-system execution guidance that results in the end product, the engagement of the space capability. In this view, the under secretary needs a small organization to integrate space-system needs or requirements that the U.S. Strategic Command (USSTRATCOM) and the various COCOMs have developed. Each COCOM is responsible for estimating and prioritizing its space-system needs and requirements and passing these to USSTRATCOM, which then passes them to the Under Secretary of

---

[8]   AFI 38-101 outlines this organizational structure.

Figure 2.4
Using the Strategies-to-Tasks Framework to View AFSPC as a Supply-Side Organization

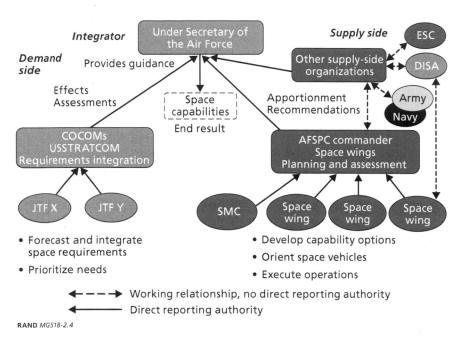

RAND *MG518-2.4*

the Air Force. On the demand side at the COCOM level, a J3/5 organization could gather these space-system needs and requirements and prioritize them. This organization is responsible for working with joint task force commanders to forecast and integrate space-system needs and requirements.[9]

Also, from this execution-level view, AFSPC and its space wings are the supply-side organizations responsible for providing space assets to COCOMs and others. These responsibilities include configuring space assets to meet needs, transmitting needed information to users, establishing schedules for meeting user needs, and overseeing execution operations. As illustrated in the figure, there are other providers of space capabilities, and the Under Secretary of the Air Force may choose

---

[9] See Tripp et al. (2006) for a theater distribution system analysis that outlines proposed J3/5 organizational roles and responsibilities.

who should supply the needed capabilities. It is also important to note that the Defense Information Systems Agency (DISA) is an integral part of the supply side because of the reliance of space missions on ground communications.

One feature of supply and demand relationships is that they are often nested, both in and outside the command or service. Supply, demand, and integrator roles are defined not only at the execution level (as in Figure 2.4) but also at other levels. An organization may be a demand-side organization at one level and a supply-side organization at another. For example, Figure 2.5 shows individual supply-side organizations (such as contractor support, Air Force squadrons, and the Air Force Materiel Command [AFMC]) and individual demand-side organizations (such as launch and operations groups) when viewed from a

Figure 2.5
The Nested Supply and Demand Relationships Within a Space Wing: Supply, Demand, and Integrator Roles

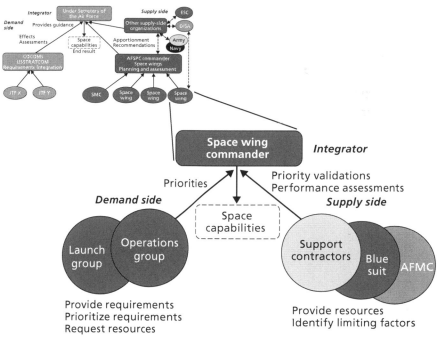

RAND *MG518-2.5*

wing commander's perspective. At the execution level, these organizations are all supply-side organizations, supplying the Under Secretary of the Air Force with the needed space capabilities. The nesting that exists in space operations planning and execution adds another layer of complexity to the overall space system.

To this point, we have introduced a framework and outlined the analytic approach. Next, we will use our framework to identify shortfalls in existing (AS-IS) policies relating to philosophies, processes, training, doctrine, systems and tools, and organizational structure. We then discuss TO-BE suggestions to mitigate these shortfalls in each area.

CHAPTER THREE

Space-System Sustainment Philosophy and Process: AS-IS Shortfalls and TO-BE Improvement Options

The AS-IS

We used the principles and philosophy inherent in the strategies-to-tasks framework to guide our evaluation of space-system sustainment. The balance between supply-side and demand-side requirements has been recognized at the HQ Air Force level, and organizations and processes have been partially developed to leverage the tension between demand and supply requirements. In the POM arena, AF/IL on the Air Staff represents the supply side; AF/XO represents the demand side; and AF/XP plays the neutral integrator role for the Chief and Vice Chief of Staff.

At HQ AFSPC, the separation of A4S (the supply side) and A3 (the demand side) is an example of applying the philosophy implicitly. Currently, A4S is responsible for developing sustainment policies and practices, but there is no strong neutral integrator. Also, no processes exist to ensure that adequate funding is received through the POM for sustainment (the integrator's analytic capability is currently provided by Plans and Programs [A8A9]). Finally, there is no explicit philosophy that guides AFSPC AS-IS support system design and development.

Another example of implicitly applying the strategies-to-tasks philosophy is at the Western Range. Currently, the Western Range support contracts have been consolidated into four main contracts (see Appendix D for a summary of these). Although these four contracts still have combined operation and maintenance (O&M) management, the Western Range's management squadron has discussed separating operations and sustainment contracts[1]—which would adhere to the strategies-to-tasks philosophy.

However, space-support doctrine is minimal. There is no explicit recognition of the natural tension between supply- and demand-side functions, and mechanisms to formally deal with the tensions are not fully developed. Few integrator processes are defined in doctrine and regulations, and supply-side processes are not fully developed. There is no recognition of the importance of the integrator role to provide guidance and checks and balances between supply and demand functions. In addition, force-development activities associated with support personnel (supply-side) do not receive the same attention as those associated with the operations (demand-side) community.

The space wings have not fully adopted the strategies-to-tasks philosophy. Supply-side and demand-side processes are collocated in the operations and launch groups (except in 21 SW, which has consolidated some maintenance functions into a maintenance group—see Appendix G for more details about 21 MXG).

Embedding supply processes within the demand-side organization can create imbalances that favor current readiness at the expense of future readiness. Balancing supply and demand processes is critical to preserving and protecting space capabilities.

Figure 3.1 illustrates the challenges of balancing current and future capabilities. Dahlman and Thaler (2000) highlighted this balance, describing the issue this way:

> The official DoD dictionary defines *operational readiness* as "the capability of a unit/formation, ship, weapon system or equipment to perform the missions or functions for which it is organized or designed." A distinguishing feature of the approach taken is that this concept is applied to both peacetime and wartime tasking.

---

[1] Discussions with 30 Range Management Office (RMO/RMS) personnel, August 2005.

Figure 3.1
Balancing Supply and Demand Is Central to Preserving and Protecting
Capabilities Now and in the Future

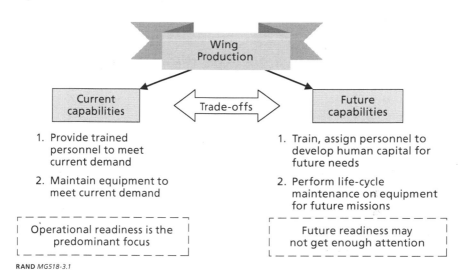

On the most basic level, U.S. Air Force wings and squadrons are designed to produce two overarching and intimately connected outputs related to readiness. The first is the ability to provide current military capabilities, i.e., the activities universally associated with operational readiness. If a wing had to go to war today, how well would its capabilities match up with the demands of the COCOMs? Are the right numbers of personnel trained appropriately? Is equipment in good working condition, with an adequate level of supplies? Can the requisite number of effective sorties be generated?

The second is the current production of future capabilities. While this usually receives less attention, it is equally important. We emphasize these activities here precisely because they tend *not* to be emphasized in actual planning and programming. DoD and U.S. Air Force guidance on and management of readiness traditionally emphasizes operational readiness, and the requirements for maintaining this readiness are explicit. The production of future capabilities, through the rejuvenation of human capital by formal and on-the-job training, is not normally recognized as an equally important tasking. It is a capability that is assumed to be embedded in units but that often is not. Units deployed to support contingency operations often postpone building future capabilities to provide current ones. The longer this continues, the more future commanders will be limited by having a less-experienced, less-capable force from which to draw (Dahlman and Thaler, 2000).

Wings have two competing taskings or outputs: current and future capabilities. Today, attention is predominantly focused on current capabilities. Future readiness may not receive enough attention.

Both the space broad-area review (NASA Office of Logic Design, 1999) and the Space Commission report (2001) recognized the need for improvement in space management and engineering processes. The Space Commission (2001, p. 9)

> concluded that a number of disparate space activities should promptly be merged, chains of command adjusted, lines of communication opened and policies modified to achieve greater responsibility and accountability. Only then can the necessary trade-offs be made, the appropriate priorities be established and the opportunities for improving U.S. military and intelligence capabilities be realized.

NASA Office of Logic Design (1999) called for an integrated approach to space processes to focus on mission success (for example, space wings have integrated acquisition, operations, and maintenance). Although we agree conceptually that processes should be integrated, the expanded strategies-to-tasks framework provides a philosophy and methodology for balancing operational requirements with sustainment needs by separating supply-side processes from demand-side processes. The strategies-to-tasks framework also provides a specific role for a neutral integrator: weighing trade-offs between operations and sustainment. Even though the processes have been separated using the strategies-to-tasks framework, the overall focus is still on mission success.

Implications

Supply, demand, and neutral integrator principles should be recognized explicitly, and doctrine, processes, and organizations should be refined to apply them with more rigor. Roles should be explicitly defined, stressing the importance of all three aspects of the strategies-to-tasks framework.

Besides the intended creation of tension between supply and demand processes, the strategies-to-tasks framework underscores the need for analysis to provide the integrator with information on the effects of supply decisions and the effects of operational decisions. A focus on supply and demand metrics needs to be encouraged and can lead to improvements in metrics from the demand, supply, and integrator perspectives. For example, the integrator may need an analytic arm to weigh demand- and supply-side inputs and provide a neutral viewpoint.

The explicit adoption of the strategies-to-tasks framework should also help all involved in the sustainment of space systems to understand the philosophical underpinnings of the AFSPC organization and ensure that these changes are strengthened over time and survive changes in leadership. Doctrine could be enhanced to recognize that the HQ AFSPC organizational structure—along with the combat wing structure—does follow the strategies-to-tasks philosophy and guiding principles and to outline why it is beneficial to organize according to these principles.

Understanding the benefits of adopting the philosophy is important. If the benefits are explicitly recognized, organizational structure may not be as important. However, to fully realize the benefits of the philosophy based on the strategies-to-tasks framework, the philosophy should drive development of both the organizational structure and the metrics.

We now turn our attention toward processes that follow the strategies-to-tasks philosophy and framework that could enhance sustainment management if set in place.

Space-System Sustainment Processes

Recently, AFSPC has made some strides in improving support processes. Enhancements have been made in the quality assurance (QA) program in understanding how to assess contractor processes to perform necessary sustainment work and the management information system associated with QA.

Both 30 Range Management Squadron, Quality Assurance Flight (RMS/RMQ) and 45 RMS/RMQ QA offices are charged with ensuring that contractor performance, cost, schedule, products, and services conform to requirements in the Federal Acquisition Regulations, 63-series Air Force instructions (AFIs), and contract statements of work (or statements of objectives). Their approach balances insight (proactive and prevention focused) with oversight (detection and reaction focused).

The 30 SW QA program reorganized and assigned QA specialists to specific processes, such as concept development, product and service delivery, and system retirement. This functional orientation provided much better insight into contractor processes. The 30th also provides traditional oversight inspections of services and products, but the inspections are now related to the extent that the contractor has mature and repeatable processes. If the contractor has mature and repeatable processes, less time is spent checking the outputs of the process; if the process is not mature, more time is spent checking the outputs (Walker, 2005).

To facilitate data collection, the 30 RMS/RMQ QA office developed an innovative QA management information system using Microsoft Access. The management information system provides data reduction via populated forms, user-defined reports, and graphical representations. Users input contract surveillance data and observations that can be compiled into reports showing indicators and identifying trends. Functional commanders in the wing and SMC program managers use these management information system reports, which also serve as input to HQ AFSPC quarterly sustainment reviews (QSRs). A future goal is to create an automated interface between legacy databases (such as the Core Automated Maintenance System [CAMS]) and contractor databases to feed reliability and maintainability metrics into the management information system and QSR deliverables.

The 45 RMS/RMQ QA office has a similar QA management information system that uses the Clarion software platform. Although the two wings have shared information, each is satisfied using its own product. While both systems perform adequately, the justification for separate systems is questionable. For long-term sustainment, AFSPC may wish to adopt a standardized, open-architecture management information system operating on a common platform tailorable to each wing's requirements so that data may be exchanged and so that there is only one system to update and modify. Eventually, a standardized management information system could even span changes in contractor organizations.

In addition to QA, Call Henry, Inc., which services the Launch Operations Support Contract at Vandenberg AFB, has applied RCM concepts innovatively to range infrastructure and equipment (see Appendix D for a summary of range management contracts). RCM is a maintenance concept that is steadily replacing the concept of repairing or replacing equipment either when it fails or after a predetermined number of operating hours or cycles. RCM is a proactive approach that continuously evaluates the condition of equipment to achieve maximum effectiveness at reduced cost.[2]

In an approach very similar to operational risk management, Call Henry has a robust method for identifying and prioritizing sustainment actions competing for limited resources. For each critical equipment item and facility, the company analyzes the failure mode effects and criticality to assess the likelihood of failure and the severity of the consequences—how it affects the mission. Plotted two-dimensionally, this gives a graphical representation of risk, providing the government-

---

[2] RCM is an application of a larger concept known as *condition-based maintenance* (CBM), also referred to in the literature as "CBM+". In November 2002, then–Deputy Under Secretary of Defense for Logistics and Materiel Readiness, Diane K. Morales, issued a policy letter to DoD and the military services to expand implementation of CBM+. According to the *Defense Acquisition Guidebook* (Defense Acquisition University, 2003, para. 5.2.1.2), the goal of CBM is to perform maintenance only on evidence of need and includes "appropriate use of diagnostics and prognostics through the application of RCM." The Assistant Deputy Under Secretary of Defense for Maintenance Policy, Programs, and Resources (2004, p. 2) confirmed that the CBM+ focus, as one of the six initiatives under Future Logistics Enterprise (FLE), is to "include new acquisition and, where cost effective, legacy weapon systems that are maintained in the organic and commercial sectors . . . to increase operational availability and readiness throughout the weapon system life cycle at a reduced cost."

contractor team insights into when repair or replacement should occur and how often to apply RCM techniques (see Figure 3.2). This deliberate, analytic approach aids prioritization of the workload so that the government-contractor team can base its assessments and implementation decisions on how to best use available resources.

Examples of RCM techniques include the following:

- battery impedance measurement, which permits replacing individual batteries that have reached the end of their service life, rather than all batteries
- motor vibration analysis to detect worn bearings, allowing them to be replaced before failure occurs
- oil analysis to detect deterioration of oil-wetted components
- infrared thermographic detection to locate faults in electrical, heating, ventilation, and air conditioning systems
- ultrasonic leak detection.

Call Henry also uses RCM techniques for maintaining facilities, including corrosion control on exterior surfaces (Call Henry, 2005). The 45 RMS Launch Operations Support Contract contractor is also implementing RCM. Appendix E provides a facility corrosion-control example.

AFSPC has also enhanced processes for tracking system component performance and sustainment requirements and for briefing the results to senior leaders on a continuing basis. This process, called the QSR, is used to assess the sustainability of space systems supported by ESC and SMC.

The QSR is moving away from reporting what has happened to predicting what might happen. The process and the focus on sustainment metrics are noteworthy efforts. The QSR is also identifying resource requirements needed to sustain space systems and trying to predict when shortfalls will occur. The HQ AFSPC Maintenance Management Section developed and refined performance standards for various system performance and health measurements, when they had not previously been developed. It also developed a way to automate

Figure 3.2
Call Henry's Streamlined RCM—Action Step

Likelihood	1	2	3	4	5	Legend
5	D	B	A	A	A	**A** High: Inspection and immediate repairs required.
4	D	C	B	A	A	**B** Medium high: Sampling and RCM recommended; repair in next 1 to 3 months
3	D	C	C	B	A	**C** Medium low: Long-term possible repair candidate; add to RCM to keep status from degrading.
2	D	C	C	B	B	**D** Low: Equipment is sound, just not so new. Risk is low; RCM sparingly
1	E	D	D	D	C	**E** Negligible: Probable run-to-failure candidate.

Consequences

SOURCE: Adapted from Call Henry, Inc. (2005, slide 12).
RAND *MG518-3.2*

parts of the processes of developing data and briefings to establish process control and to ensure standardized presentations from the various systems. In addition, it developed the algorithm to create a system measurement called Logistics Capability (commonly referred to as LOGCAP), which displays logistics capability in a form and format acceptable to the A3 and which moves toward a leading rather than lagging indicator of system health.

While draft AFI 21-203 mandates the QSR process, it currently lacks governing documents. Over the past several years, the focus has been on development and execution of the QSR, not on documentation. AFSPC is currently working to develop some governing documents.

Furthermore, significant efforts have been made to aggregate support contracts by site rather than by individual space system. AFSPC has taken several steps to consolidate common space-system ground systems and infrastructure-support contracts. In the past, each system may have had a separate contract that addressed only a portion of the ground equipment at a site, including general maintenance of that piece of equipment. As part of the effort to consolidate, each unique piece of equipment dedicated to specific systems may still have a separate contract, but all have general maintenance requirements (for example, power systems and corrosion control) that are combined in one contract. Consolidating these contracts appears to have yielded economies of operation.

Another example of a process enhancement is the use of the Integrated Priority List (IPL) for AFSPC oversight of range modifications. The 30 and 45 SWs use the IPL to identify, validate, approve, control, prioritize, and monitor status of launch and test-range architecture modifications. Governed by AFSPC Instruction 21-104 and local instructions, the IPL is managed by AFSPC/DRS and the range management squadrons. The Range System Program Office and various wing agencies (XP, SE, ROPS) serve as advisors to the range management squadron for technical assessments, planning, safety, and operational factors in the development of an initial IPL score. Wing-level requirements validation boards review and validate requirements scores and, for projects over $500,000, forward recommendations to HQ AFSPC. Score weighting is the sum of the wing, SMC, and HQ AFSPC factors shown in Table 3.1. For approved initiatives, SMC determines a technical solution, creates an engineering change proposal, and obtains configuration control board approval. The IPL is then used to monitor progress of funded and approved initiatives. IPL requirements can be generated by any organization on the range using an online electronic tool to capture the requirement.

AFSPC/A4S also has focused attention on metrics. A draft sustainment metrics handbook has been prepared and circulated to all AFSPC organizations for use in the QSR process. AFI 21-203 provides high-level guidance on metrics.

Even though much has been done to improve sustainment processes in AFSPC, more can be done. First, space sustainment planning and execution processes do not relate how alternative sustainment plans and expenditures affect joint operational effects. Feedback loops and diagnostics have not been established that relate performance measures of effectiveness to the parameters needed to achieve the desired

Table 3.1
IPL Requirements Scoring Factors

Level	Requirement	Score
Wing	Range mission (general issues)	0.0800
	Range mission (waiver impact)	0.0533
	User mission	0.0576
	Work-arounds	0.0576
	Infrastructure	0.0404
	O&M reliability, maintainability, and availability (RMA) (spares)	0.0370
	O&M RMA (problem item fails)	0.0370
	O&M RMA (failures until "red")	0.0370
	Wing total	0.4000
SMC	Sustainment RMA	0.1286
	Contract impact	0.0857
	Cost savings	0.0857
	SMC total	0.3000
HQ AFSPC	Other impacts	0.3000

operational effects (see Figure 3.3 for a sample closed-loop feedback system). The "public good" aspect of sustainment funding is not fully understood (for example, GPS sustainment funding decisions and their effects on gross domestic product). Funding could perhaps be increased if the relationship between sustainment and joint effects (commercial consequences as well) were more clearly established and used in the budget process.

Furthermore, some current financial management policies specify basing future sustainment requirements on projecting past expenditures. Therefore, estimation of sustainment requirements is limited to straight-line projections with no recognition of the effects of aging on systems and increasing sustainment costs. These policies also negate the benefits that could accrue from applying RCM procedures.

Figure 3.3
Closed-Loop Planning and Execution Process

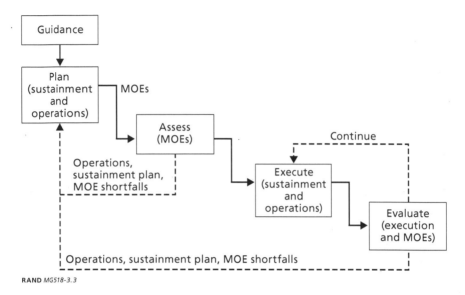

RAND *MG518-3.3*

Since supply-side processes are embedded in demand-side organizations, the impression is that meeting immediate operational requirements gets more emphasis than providing sustainment to ensure capabilities are available in the future.[3]

System acquisition processes (for example, those SMC and ESC handle) can be integrated better with ongoing sustainment processes. For instance, during technology insertions in the ranges, "real" systems are used for conducting technology demonstrations. Using simulators instead may be advantageous for modernizing and testing upgrade

---

[3] Discussions with 30 RMS, 45 RMS, 21 OG, and 460 OG. With the importance of space operations, difficulties in current operations were often overcome by individuals using work-arounds in processes, which eliminated data on the balance current between operations and sustainment. In addition, metrics and data showing the effects of focusing on current operations to the detriment of sustainment are often not tracked or even established.

components. This would eliminate the sustainment effects of testing upgrades on the live system.

In addition, the integrator's responsibility to balance capabilities now and for the future has not been fully recognized, and periodic formal evaluations of potential trade-offs between demand and supply should be required. For example, scheduling time for sustainment actions needs to be given more attention. While acquisition, launch, and operations schedules are shared with the range, schedule slippages often affect range maintenance activities. To address these slippages, range sustainment managers need to develop agile maintenance schedules and practices that can be modified if dynamic schedule slippages change range requirements. Additionally, coordination of future requirements to support acquisition at the ranges could be improved.

Finally, the current set of Air Force sustainment processes does not include some key processes and players that affect space-system sustainment. For instance, DISA is not represented in space-system sustainment reviews, yet DISA provides significant long-haul land communications segments used in operating space systems.

The Sustainment Process—Recommended Actions

Several initiatives can be taken to address the current process shortfalls outlined above. First, the 30 SW QA process and supporting management information system is a "best practice" in AFSPC. Other space wings could develop QA processes based on the 30 SW model. In addition, the QA management information system could be expanded to include system performance metrics, as well as the contract performance metrics it already contains. Each support unit could develop specific metrics appropriate for its use, and the QA database could be populated with this information by electronic transfer of requirements from existing legacy databases (for example, CAMS/Integrated Maintenance Data System [IMDS] and Standard Base Supply System) or by direct feed from contractor databases, provided contract data requirements lists were adopted to do this. Once this data transfer is complete, the QSR data could be extracted from this database automatically.

Second, RCM practices should be expanded within AFSPC beyond infrastructure-type equipment to include primary mission equipment.

Again, 30 SW's experience in this area can provide the other wings with a model that could be tailored to their specific needs.

The IPL used at the ranges could be expanded and implemented at nonrange wings. The process should be formalized to aid wings in identifying, validating, approving, controlling, prioritizing, and monitoring the status of upgrades and modifications to space systems.

In addition, a process for prioritizing and tracking sustainment resource requirements across the command (like a "super IPL") could be established. This super IPL could track all funding, including the sustainment money received from other agencies.

Finally, metrics need to be developed that relate sustainment resource needs to operational effects. Current metrics address individual system components and support with respect to that system. Metrics have been developed to track all components of GPS, including the Master Control Station, four ground antennas, and five monitoring stations. Metrics have also been created to track the performance of the space segment, which currently consists of 29 satellites. However, these metrics reflect past performance or effects or present report-card types of data. They do not provide predictive or leading indications of future issues that may arise in the GPS system.

The metrics specified in AFI 10-602, Attachment 9, for space systems include operational availability, operational dependability, mission reliability, logistics reliability, and mean repair time. As an example, the GPS QSR reviews operational availability, operational dependability, mean time between critical failures, mean time to repair (MTTR), mission capability, depot-level maintenance (both emergency and unscheduled), maintenance drivers, and supply drivers. Most, if not all, of these are lagging indicators, which provide a historical perspective focused on effects.[4]

On the other hand, leading indicators, such as break rate, repair rate, repeat and recur rate, scheduling effectiveness, and delayed and deferred discrepancy rates, demonstrate causes of the lagging indicator effects and thus offer a predictive form of analysis for maintenance managers. For example, the GPS program categorizes minor discrepancies under the "amber" system status. GPS program managers do not formally track amber job rates. An accumulation of these could point to a systemic problem that might otherwise go unaddressed. Mission-limiting (red) discrepancies are the main focus.

Although some systems may use leading indicators, these indicators are not formally tracked. Implementation of leading indicators may require the contractor to collect additional data and, potentially, to modify its information systems. However, CAMS/REMIS users have ready access to data and analysis on leading indicators, which are used throughout the aircraft maintenance community. But not all contractors are using the same data-collection systems. This may not pose a problem if metrics are clearly defined across the command, which would make the data-collection medium secondary to the metric.

Figure 3.4 shows, conceptually, how sustainment resources can be related to operational effects. The example in this figure shows how an analysis of ground segment support might affect GPS accuracy.[5]

As shown at the top of the figure, maintenance spending can be expected to affect equipment failure (mean time between failures

---

[4] AFPD 10-12, Space, specifies metrics for space-system readiness, satellite launch success, and on-orbit success; AFI 10-602, *Determining Mission Capability and Supportability Requirements*, specifies metrics for operational availability, operational dependability, mission reliability, logistics reliability, and mean repair time; AFI 21-108, Maintenance Management of Space Systems, specifies use of CAMS/Reliability and Maintainability Information System (REMIS) and exceptions; AFI 21-203, *Space Maintenance Management* (Draft), provides general guidance on use of management information system and metrics; and Guidelines for Reliability, Maintainability, and Availability (RMA) Metrics for the Launch Test Range System (LTRS) (Revision). Other sources may exist, particularly at the unit level.

[5] Further work in this area can be found in Snyder and Mills (2007). This report outlines criteria for analyzing how sustainment investments affect the operational performance of space systems, focusing on the GPS.

[MTBF]) and repair times (MTTR). Various scenarios could be created to illustrate the effects of adding a monitoring capability, such as the National Geospatial-Intelligence Agency (NGA) monitoring stations or use of the Air Force Satellite Control Network (AFSCN) capabilities. In addition, the scenarios could show the effects of subtracting a capability, such as a degraded monitoring or upload capability that may result from tropical storms or terrorist attacks. These combinations will affect total ground-station availability.

Ground-station availability, as shown in the middle of the figure, can affect satellite visibility in various regions. This, in turn, can be adjusted to some extent by modifying satellite positions. These combi-

Figure 3.4
Metrics and Analyses That Relate Each System Component to End-to-End Capability

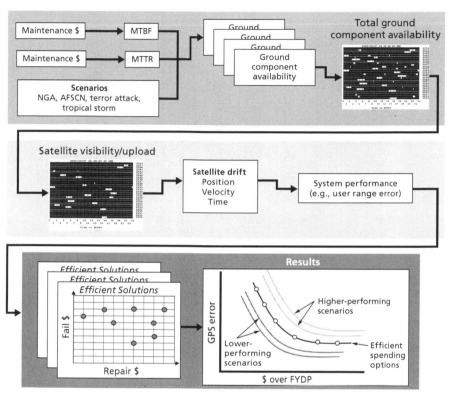

nations provide a given system performance. Finally, as shown at the bottom of the figure, various sustainment resourcing options will yield differing GPS errors over various regions.

The ability to relate support investments to system performance (effects) can be expected to produce better support resource-allocation decisions within and among space programs and to provide a quantitative rationale for support programming needs.

From a longer-term point of view, SMC, an acquisition organization that has been realigned under AFSPC, is responsible with AFSPC for developing and modifying space systems to meet the needs of the

COCOMs and other government agencies. SMC and its program offices need to specify sustainment trade-offs as they formulate their development and modification plans. ESC, which develops some space-system capabilities, does not report directly to AFSPC leadership, such as in the execution environment with DISA. For ESC-developed systems, AFSPC needs to coordinate requirements effectively with ESC and to ensure that sustainment needs are considered in development and modification of systems. The collocation of SMC and ESC elements at HQ AFSPC helps ensure these needs.

Implications

Our analysis indicates that using a strategies-to-tasks framework and philosophy to separate supply, demand, and integrator processes to improve the effectiveness and efficiency of space-system sustainment is a good first step. Once these processes are separated, many other improvements can be made. For example, the role of analysis—and informed analyses—can be emphasized when resources are constrained and tough choices have to be made. These analyses can be expected to improve understanding of the relationships between control-segment and space-segment performance. Over time, metrics can be developed to ensure informed decisions that will take into account control-segment performance, capture details appropriate for the specific audience, capture user-oriented effects, and be meaningful to a wide audience.

CHAPTER FOUR

Force Development, Doctrine, and Information Systems and Tools

This chapter addresses trends and potential shortfalls in current force development for space-system sustainment, potential doctrine enhancements, and the information systems and tools that are used to help manage space-system sustainment activities.

Force Development

Development of civilian, officer, and enlisted logistics and communications leaders with space experience is essential to the success of the AFSPC mission. While major strides have been made in improving personnel development and management for AFSPC officers, noncommissioned officers, and civilians, using the strategies-to-tasks framework would seem to make further enhancements in force development for support professionals possible.

The strategies-to-tasks framework would suggest that maintenance be managed from supply-side organizations. This structure would provide a clear career path for support management and growth, a source for advocating career development, and potential advancement up to commanding the maintenance group. This supply-side group of senior officers who eventually become maintenance group commanders could be an effective voice for ensuring that long-term training and force development take place.

Implications

Missile Maintenance (21M) and Communications Maintenance (33S) are the primary officer career fields required to supervise space-system logistics areas.[1] Although several communications maintenance officers interviewed perceive that back-to-back assignments in AFSPC are discouraged, the force development office indicated that a new program is being established with a developmental identifier to track space expertise as a step toward vectoring officers to AFSPC positions at appropriate points in their careers. The missile maintenance officer force-development team is pursuing career advancement focused on the munitions and aircraft and the ICBM maintenance areas, due to the absence of maintenance groups in AFSPC.[2] If additional maintenance groups are established in AFSPC, in keeping with the strategies-to-tasks philosophy, 33S and 21M force-development teams should modify their focus to ensure continuous production of competent and experienced space-focused maintenance group commanders and deputies.[3]

Noncommissioned officer development could benefit from more-focused training in space systems to augment the present Harris-type short courses; noncommissioned officers engaged in quality assurance evaluation (QAE) activities could benefit from some additional training and education in interpreting contractual documents. The Defense Acquisition University may provide some offerings (distance learning or in person) to meet these requirements.

Space credentialing may also benefit force management. At a minimum, special experience identifiers for space systems should be developed and applied to facilitate the filling of key space jobs in the officer and enlisted ranks. A new 10-day course, taught by 392nd Training Squadron (TRS), has recently been implemented to enhance

---

[1]   Except in launch groups, in which AFSPC has limited 21M participation.

[2]   The 21 SW at Peterson AFB has a maintenance group. See Appendix C for more details.

[3]   The AFSPC Strategic Master Plan (P5.5.2) calls for training, exercise, and education for "all areas" of the space community. Staats and Abeyta (2005) recommends that the space cadre "include intelligence officers and logistics officers" because "it is easy to overlook logistics considerations during the development and integration of new doctrines, we must address those issues in advance rather than resolve them in hindsight."

the knowledge of 2M0 enlisted personnel prior to, or shortly after, assuming duties in spacelift maintenance.

Civilians present the most stable, experienced element of the AFSPC logistics workforce because of their deep knowledge and infrequent changes of assignment. Civilians too could benefit from understanding the principles associated with the strategies-to-tasks framework for sustainment. The credentialing of space logisticians could begin with the civilian workforce. By starting with civilians, any potential problems in credentialing criteria or approach could be identified before moving the training to enlisted and officer personnel.

Doctrine

Although progress has been made in expanding doctrine to address support functions, support doctrine is not as robust as operational doctrine. In addition, space support doctrine is not as mature as aircraft support doctrine. For instance, Air Force Doctrine Document (AFDD) 2-2, *Space Operations*, devotes only two paragraphs to logistics. AFDD 2-4, *Combat Support*, treats space as a unique or special case. The AFDD 2-series publications could benefit from revision to include space-centric support concepts.

Presently, AF/ILGX and ILGC are engaged with the Air Force Doctrine Center, other HQ Air Force elements, the Secretary of the Air Force, and all major commands (MAJCOMs) to develop AFDD 2-4 subpublications more fully. Four subpublications are planned for development, with at least three presently being drafted to align with current Agile Combat Support capabilities.

Implications

AFSPC can make significant contributions by defining and inserting space support doctrine into Air Force publications and USSTRATCOM support doctrine into joint publications. AFSPC could work with HQ AF/ILGX on developing new AFDD 2-4 subpublications and could identify demand, supply, and integrator roles. The development of the strategies-to-tasks framework, which outlines the philosophy for separating supply-side, demand-side, and integrator functions, would contribute to doctrine and show how aircraft, missile, and space systems can follow the same support philosophy, even if the specific implementations vary to some extent based on mission specifics, nature of systems, or history of support—organic or contractor at both the service and joint levels. An organizational structure to support the strategies-to-tasks framework may not be as important if roles and responsibilities are explicitly recognized in doctrine.

Information Tools and Systems

Not all space systems provide the same level, in depth or breadth, of maintenance data collection and system and component status reporting. Command policy requires the use of CAMS and/or IMDS for equipment items; however, particularly for facility reporting, use of systems the contractor has developed (unit-level) is common. Call Henry's 30 SW Launch Operations Support Contract project management review and status briefings demonstrate this (see Appendix E for more information).

Standardization of data collection and reporting may be less important than standardized metrics. Performance-based logistics ("pay-for-performance," specifying the "what" but not the "how"), cost-effectiveness, and commercial practices may dictate a best-value solution that differs among systems. These logistics-transformation initiatives are widespread throughout the command, particularly as major range contracts come up for recompetition and renewal. The critical issue is the ability to obtain key performance metrics so that trade-offs can be made on where to invest sustainment dollars if resources are constrained. These trade-off decisions should be based on the collection of similar information among systems so that like comparisons of metrics can be made.

In addition to metrics, sustainment actions and the effects of not performing needed sustainment actions should to be tracked. Sustainment actions and the lack or delay of them can affect both military and commercial operations that, in turn, affect joint operational effects and even the gross domestic product, in some cases. The cost of not making a space launch is tangible in addition to the operational effects of not having the space capability. Many of the typical indicators (leading and lagging) can be used to better prioritize sustainment actions, if properly integrated with acquisition schedules and operational schedules (planned). Others can be used to gauge contractor performance and award fees. To make trade-offs when resources are constrained, both measurements and metrics are needed. RCM actions should be tied to these metrics as well.

Implications

Information and system and tool guidance improvements are under way throughout the space community. Following is a list of instructions that have been created or updated to provide guidance regarding space-system tools:

- AFPD 10-12, *Space*. This document specifies metrics for space-system readiness, satellite launch success, and on-orbit success.
- AFI 10-602, *Determining Mission Capability and Supportability Requirements*. Attachment 9 specifies metrics for operational availability, operational dependability, mission reliability, logistics reliability, and mean repair time.
- AFI 21-103, *Equipment Inventory, Status, and Utilization Reporting*. This instruction describes minimum essential subsystem listings—see the AFSPC supplement.
- AFI 21-108, *Maintenance Management of Space Systems*. This instruction specifies the use of CAMS/REMIS and exceptions.
- AFI 21-116, *Maintenance Management of Communications Electronics*. This instruction provides maintenance management guidance for communications-electronics systems and equipment.
- AFI 21-203, *Space Maintenance Management* (delayed draft). This draft provides general guidance on use of management information system and metrics.
- *Guidelines for Reliability, Maintainability, and Availability* (RMA) *Metrics for the Launch Test Range System*, rev., October 2005. These

guidelines describe the process for computing metrics to achieve standard RMA metrics for the launch test range system.

- AFSPCI 10-1208, *Launch and Range Roles and Responsibilities.* This instruction establishes space launch, range O&M, and launch readiness review process roles and responsibilities.
- Technical Order 00-20-2, "Maintenance Data Documentation." This order provides policy and guidance for collection and documentation of maintenance data.

More work can also be done on developing leading indicators. These indicators should reflect the cause that created the effect and be used for prediction. Decisionmakers should be able to focus their attention on future problems by using leading indicators. (See Chapter Three for more information about leading indicators.)

CHAPTER FIVE

Space-System Sustainment AS-IS Organizational Structure and TO-BE Improvement Options

We now turn to the organizational structure currently used to support space-system sustainment processes. First, the creation of AFSPC/A4S places supply-side issues in one headquarters organization. AFSPC/A4S is the focal point for space-system sustainment issues (with the exception of space launch and sustainment funding). It provides written guidance to warfighters, internal AFSPC organizations, and supporting organizations on life-cycle sustainment matters, performance assessment, and reporting issues of concern to higher leadership for resolution. However, there is no development of sustainment (POM) requirements in A4S. Second, all operational issues are centered in AFSPC/A3. However, A3 still has some embedded supply processes. For example, A3 is the focal point for operational issues and operations and sustainment funding.

In addition, the AFSPC commander is implicitly recognized as the integrator, with strong input and analysis from AFSPC/A8A9. The AFSPC commander's role could be explicitly recognized, and the commander could require frequent formal meetings for making trade-offs between operational and sustainment needs. This format would help ensure development of appropriate metrics over time for enhancing decisions between the need to support immediate operations and the need to "take systems down" to ensure that long-term sustainment actions are made. This may not be a pressing issue at present, but as the

Operationally Responsive Space concept becomes a reality,[1] balancing between current operations and long-term sustainment to ensure future capabilities may become more difficult if processes for making difficult trade-offs are not established now.

The strategies-to-tasks framework supports the collocation of ESC and SMC personnel at AFSPC. This placement puts elements of the acquisition community, a supply-side organization, at the same location as the operator of space systems, an execution-related supply-side organization. This placement can facilitate coordination of short- and long-term sustainment actions and trade-offs.

Further, SMC has been realigned under AFSPC, putting the focal point for developing and modifying some space systems (but not those acquired and modified by ESC) at one location. This alignment can facilitate better decisionmaking that deals with concept development, acquisition logistics, day-to-day operations, depot-level maintenance, sustaining engineering, and weapon system disposals. Draft AFI 21-203 addresses the roles of SMC and AFSPC organizations in sustaining space systems throughout their life cycles.

While AFSPC implicitly adopted the strategies-to-tasks framework in organizing its headquarters elements, the wing-level organizations did not. In the five wings we visited, maintenance is embedded in the space operations squadron (50 SW), space warning squadrons (21 SW), or space launch squadrons (SLSs; 30 SW and 45 SW) or is the responsibility of SMC (460 SW). At 30 SW and 45 SW, maintenance and operations contracts are intertwined, rather than separated, and operations group personnel oversee the contractors.

Integration may occur at different levels in the different space wings today.[2] However, the existing integration may be a result of personalities rather than an explicit recognition of the balance that should be maintained between supply and demand. Doctrine and training do not explicitly emphasize the strategies-to-tasks framework. Over

---

[1] The Operational Responsive Space concept is a small-scale, single-mission capability that can be deployed at need, focusing on warfighter support.

[2] Currently, for example, in 2 SOPS, the operations officer (DO) is the demand side, the MA is the supply side, and the commander acts as the integrator.

time, with changing leadership, the organizational arrangement and placement of supply-side functions in a demand-side organization may implicitly favor supporting current operations over longer-term sustainment needs.

The wings also lack a standard support organizational approach (at least between the range wings and other wings and among other wings). For example, communications support is aligned under three different types of groups: 50 Network Operations Group (NOG); 30, 45, and 460 OG; and 21 MXG. Standardization of organization may not be as important as the adoption of the strategies-to-tasks philosophy because of the differences among space systems and their individual uniqueness. The explicit adoption of the strategies-to-tasks philosophy and framework would assign supply processes to a single organization, demand processes to another, and integrator processes to a third. These organizations can be different at each wing, depending on the mission and the systems supported, but the separation of the functions could help ensure that the capabilities of current and future systems are better balanced. The explicit recognition of the strategies-to-tasks framework should also help all involved in the sustainment of space systems understand the philosophical underpinnings of the AFSPC organization and ensure that these changes become stronger over time.

Finally, the Warfighting Headquarters Initiative plans to combine the A4 and A6 organizations at HQ AFSPC. This differs from the HQ U.S. Air Force, which has chosen to combine the A4 and A7 organizations. This difference may present additional "integration" issues between maintenance and logistics and communications, not only in AFSPC but also with the rest of the Air Force.

Sustainment Organization Structure Options

To address the organizational shortfalls mentioned above, we considered three organizational options for the space wing construct. Again, the specifics for each wing will differ depending on the specific nature of the systems the wing supports.

The first, or baseline, option is to leave the wing-level construct as it is currently organized. Option 2 leaves the processes in the same organizations as in the baseline, but improves the understanding of the strategies-to-tasks framework. Option 2A appoints an integrator within the group staff(s), while option 2B appoints an integrator within the wing staff. Both variants of option 2 enhance metrics to facilitate better short- and long-term decisionmaking. In both, the newly appointed integrator would hold frequent meetings to ensure that sustainment and current operations were balanced. Option 3 realigns sustainment tasks and places them in a maintenance organization that is on a par with the operational organization. Option 3 also recognizes the wing commander as the integrator, with the wing A8A9 supplying analytic support as needed. Option 3 also requires all those involved, including those outside AFSPC—for example, DISA—to send representatives to sustainment meetings.

We used three criteria to evaluate each of the options. The first criterion is the clarity of roles in ensuring that support is balanced to protect and preserve space capabilities now and for the future. The second is the effect on resources, specifically the number of people needed to implement the option. The third is the effect on training and other related force-development issues.

50th Space Wing Organization

First, we consider 50 SW, which provides command and control of communication, navigation, warning, and surveillance satellite weapon systems by operating and supporting such systems as GPS, the Defense Satellite Communications System, and MILSTAR and manages the worldwide AFSCN.[3] The options for this wing's organizational structure are shown in Table 5.1.

Option 2 refines the baseline structure by providing a neutral third party to create a balance between current operations and long-term fleet health and sustainment. This option is similar to a mid-

---

[3] See Appendix B for organizational options for HQ AFSPC and the 21, 30, and 460 SWs.

Table 5.1
Organizational Options for the 50th Space Wing

Option 1: Baseline (AS-IS)	Option 2: Integrator at the		Option 3: Form an MXG
	Group Level (A)	Wing Level (B)	
Maintenance activities exist in each SOPS in the OG	Create a neutral supply-demand integrator on the OG staff	Create a neutral supply-demand integrator on wing staff	Rename the NOG as MXG and divest of nonmaintenance functions
Some consolidated maintenance functions (e.g., maintenance control) exist in the NOG	Retain baseline structure in the OG and NOG	Retain baseline structure in the OG and NOG	Pull maintenance functions out of SOPS and consolidate into a space maintenance squadron (SMXS) in the MXG
NOG designated as lead for maintenance by 50 SW commander			Create an MOF

1990s modification to the objective wing, which has several similarities to the AS-IS construct currently in place at 50 SW. The objective wing's modification created the deputy operations group commander for maintenance position to serve as an integrator in early 1996. This position, typically filled by a lieutenant colonel career maintenance officer, was placed on the operations group staff to advise the operations group commander on maintenance matters, serve as liaison to the logistics group, and standardize maintenance across the flying squadrons. However, because O-level maintenance still worked for the flying squadron commanders, the new position was perceived as having responsibility for standardization and overall direction but little to no authority. A group- or wing-level integrator position in AFSPC with a clearly defined role, responsibilities, and authority could provide some improvement at minimal cost and could follow some of the principles of the strategies-to-tasks framework.

The third option, creation of a maintenance group, aligns supply and demand and integrator functions according to the strategies-to-tasks framework. It is also consistent with the combat wing organization (CWO) structure implemented in 2002 at the direction of the Chief of Staff of the Air Force.[4] In this option, the current network operations group would be redesignated as a maintenance group and divested of nonmaintenance functions. The latter would typically be reassigned to the mission support group (MSG). The space operations squadron's maintenance activities would be reassigned to space maintenance units (SMUs) in a new SMXS assigned to the maintenance group, with one space maintenance unit supporting each space operations squadron. To provide common maintenance management functions for the maintenance group, a maintenance operations flight (MOF) would be formed from resources dispersed throughout the current the network operations group, space operations squadron, and space communications squadron (SCS) and would report to the maintenance group commander.

Figure 5.1 illustrates the AS-IS organization of 50 SW. As shown, 50 SW has a number of space operations squadrons. The typical space

---

4   As set forth in HQ U.S. Air Force Program Action Directive 02-05 (2002).

operations squadron in the operations group has a maintenance unit led by a civilian and two company-grade (lieutenant or captain) communications officers and is staffed with communications maintenance noncommissioned officers. These maintenance units oversee system status and contractor maintenance activities. They also plan and execute modifications and support current operations. Responsibilities for quarterly remote-site QAE visits are generally not shared among space operations squadrons despite collocation of equipment. Each SOPS DO approves schedule requests for coordinating equipment downtime independently and chairs modification boards. The DO works directly for the SOPS commander, as second in command.

The wing commander has charged 50 NOG with overall responsibility for space-system maintenance. The group has direct authority over only the communications maintenance and related functions in 850 and 50 SCS and the three space operations squadrons in the group. The two communication squadrons are responsible, respectively, for

Figure 5.1
An AS-IS Graphical Representation of the 50th Space Wing

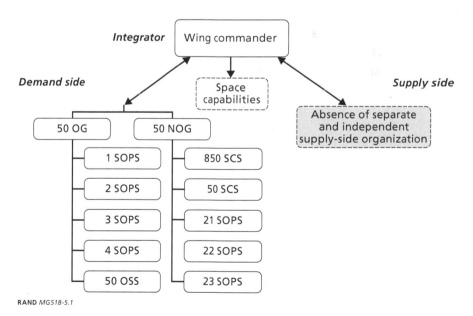

common direct mission support and general communications support to 50 NOG and other wing organizations.[5]

Figures 5.2 and 5.3 show a strategies-to-tasks representation for options 2A and 2B, respectively. In Figure 5.2, the integrator is at the group level. In Figure 5.3, the integrator is at the wing level.

Figure 5.4 shows a strategies-to-tasks representation for option 3. Here, the wing commander is the integrator, with staff support, as needed, from the wing A8A9.

In this option, the network operations group forms the basis for the maintenance group. The embedded maintenance units in the current space operations squadron form the basis for the space mainte-

Figure 5.2
A Strategies-to-Tasks View of Option 2A for the 50th Space Wing

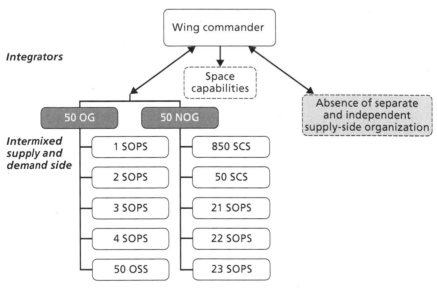

NOTE: In terms of our analysis, the SCS merger in February 2006 would have minimal effect if the recommendation to realign non–direct-mission support to the MSG were followed.
RAND MG518-5.2

_____

[5] In February 2006, 850 SCS at Schriever AFB was deactivated, and its missions and resources merged with 50 SCS, which had been providing both direct mission support and indirect (base-level services) support to 50 SW.

Figure 5.3
A Strategies-to-Tasks View of Option 2B for the 50th Space Wing

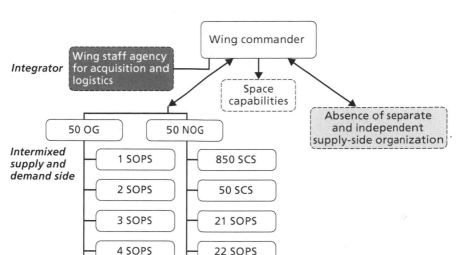

NOTE: In terms of our analysis, the SCS merger in February 2006 would have minimal effect if the recommendation to realign non–direct-mission support to the MSG were followed.
RAND MG518-5.3

nance squadron's space maintenance units.[6] One space maintenance unit may service multiple space operations squadrons, as is proposed for the former the network operations group's space operations squadron (21, 22, and 23 SOPS). The SMXS has a squadron commander, command section, and maintenance officer responsible for managing the production effort and standardizing processes across the space maintenance units.

To enhance the maintenance group's focus on core maintenance requirements, the newly formed group would realign 50 SCS, with its flights providing communications that are not directly related to the mission (base local-area network, telephones, audiovisual, etc.) to

---

[6] Demand-side activities in the SOP would remain under the operations group. If personnel are performing both supply- and demand-side activities, they would be placed in the organization with the preponderance of their activities.

Figure 5.4
A Strategies-to-Tasks View of Option 3 for the 50th Space Wing

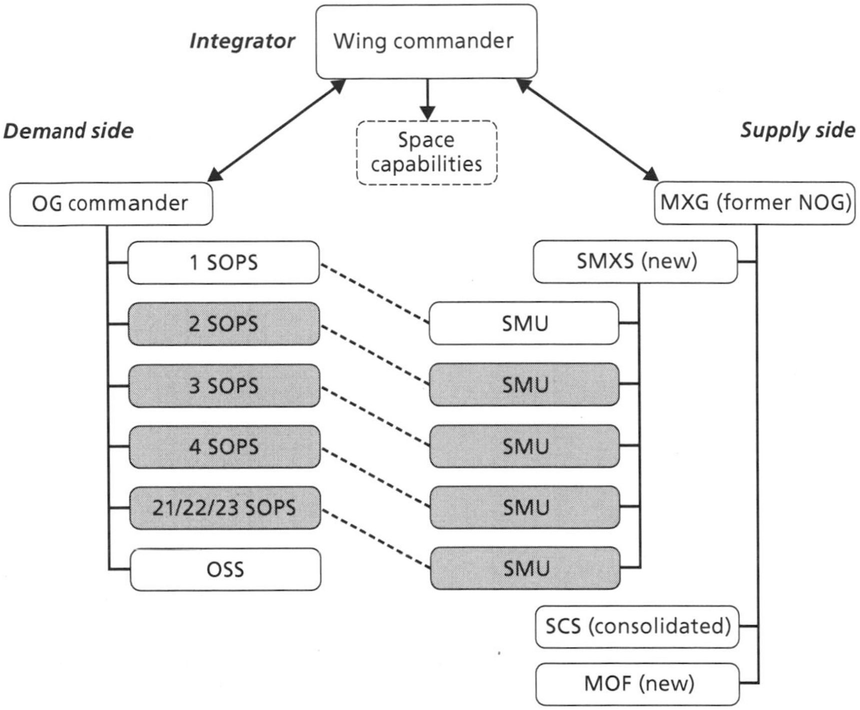

NOTES: If the entire 50 SCS were retained in a new maintenance group (as is the case in 21 SW), the size of the proposed maintenance group would increase by 81 percent to 401 authorized positions. See Appendix C for the manpower calculations.

MOF includes functions for Training Management, Maintenance Data Systems Analysis, Scheduling, and Quality Assurance.

RAND MG518-5.4

50 MSG. All generalized (common to two or more space operations squadrons) direct mission communications would be consolidated in the 850 SCS.

To provide common maintenance management functions for the maintenance group, a maintenance operations flight would be formed from resources now dispersed throughout the current network operations group, space operations squadron, and space communications squadron. The maintenance operations flight would report to the maintenance group commander. Maintenance operations flight functions are integrated scheduling, training management, analysis, supply liaison, QA, and so forth. The 21 MXG MOF serves as an example (see Appendix E).

Table 5.2 summarizes our analysis for 50 SW. The far left column shows the evaluation criteria we used to compare options. The second, third, and fourth columns show how each option rates with respect to each of these criteria.

Regarding the first criterion, we observed high levels of synergy and responsiveness to current operations in the baseline structure. The cross-training of maintenance and operations personnel that occurs on the operations floor is beneficial for developing future integrators, necessary in the strategies-to-tasks framework. However, the responsiveness was highly reactive rather than proactive. Option 2 is expected to sustain the synergy and responsiveness to current operations and to foster broader oversight at the group or wing level. Option 3 sustains high synergy and responsiveness to current operations but also improves focus on longer-term sustainment issues, following a "break-in" period of organizational adjustment, as most flying wings experienced during the CWO restructuring.

Concerning sustainment issues—future system readiness—we learned that the baseline structure commonly permits delays in scheduled maintenance and modifications, particularly for GETS requests. These requests are assessed by 2 SOPS in briefings emphasizing pros and cons and forwarded through 50 SW to 14AF for a decision. A broader O-6–level MXG view of GETS effects (and similar functions of other space operations squadrons) could improve management and permit more timely execution of sustainment activities. In addition, broader oversight of requirements screening panels and modification control boards could improve focus on the long-term health of key systems.

Mentorship of Communications (33S) and Maintenance (21M) officers, particularly company-grade officers, would be expected to improve if an O-6 position (for example, a maintenance group commander) were created with direct authority or responsibility for all wing-level maintenance, as occurred with the institution of the CWO.

Table 5.2
Summary Evaluation of 50th Space Wing Organizational Options

Evaluation Criteria	Option 1: Baseline (AS-IS)	Option 2: Integrator at the		Option 3: Form an MXG
		Group Level (A)	Wing Level (B)	
Operations-maintenance synergy and responsiveness to current operational requirements	High synergy, but reactive	High synergy Potential to be proactive	High synergy Potential to be proactive	High synergy Proactive
Fleet-health focus	No senior-level advocacy Delays in scheduled maintenance and modifications common	Some improvement contingent on level of authority assigned to integrator	Some improvement contingent on level of authority assigned to integrator Has greater potential across the wing	Senior-level (O-6) advocacy Improved balance of long-term fleet health and current operations
Mentorship of space supply-side, communications, and maintenance personnel	Limited	Limited	Limited	Improved by providing doctoral-level experts at MXG
Resources required	None additional	Integrator positions Some duplication is necessary	Integrator positions	SMXS overhead positions,[a] integrator positions, office space, office equipment

[a] Sourced from manpower savings accrued through consolidation.

Finally, the resources required to implement options 2 and 3 are minimal. Option 2 would require additional manpower positions for the integration organization. The organization could have a five-person staff, consisting of 33S (Communications), 63A (Program Management), and 13S (Operator) officers and 2E1 (Satellite Communications) and 1C6 (Operator) enlisted positions. This study did not identify an immediate source for these positions, which span the major functional areas requiring integration. Personnel assigned to the integration shop should be highly experienced. Civilians could fill at least two of the five positions, which would provide continuity through the changes of assignment of active duty personnel. The people who occupy these positions also need to be analysts and could require, at minimum, a master of science in operational science from the Air Force Institute of Technology.

Option 3 would require realigning positions accrued through consolidation to create the SMXS overhead. The 50 SW unit manpower documents (UMDs) and Air Force manpower standards for command staffs were analyzed to determine manpower requirements for option 3. The existing network operations group command staff structure would transfer to the new maintenance group. Sixteen positions in 850 SCS would be realigned to create a new maintenance operations flight that would be administratively assigned to 850 SCS but would report directly to the maintenance group commander. Thirteen positions from 50 SCS would be reassigned to 850 SCS to consolidate direct mission support in the latter. A 99-person SMXS would be created from the maintenance positions in the current space operations squadron. Its staff will be built from six positions recouped from consolidating all maintenance under one unit. Some grade and/or Air Force specialty code (AFSC) adjustments may be necessary. The new 50 MXG would have just over 220 positions. (See Appendix C for more details.) An Air Staff planning factor of 5-percent savings for consolidation efforts (recently used during Base Realignment and Closure scenario development), applied to the predicted size of the SMXS, would fulfill the overhead requirements without creating an additional manpower bill. This option may require beefing up the wing A8A9 organization to support integrator functions. There is no immediate source for these positions. These positions will be similar to those identified in option 2.

45th Space Wing Organization

The 45 SW provides command and control of the Eastern Range capabilities. Table 5.3 presents the three organizational options for the Eastern Range organization (45 SW).[7] The baseline structure, in which the operations group and launch control group (LCG) manage maintenance, is tightly integrated with operations and is highly responsive, although the groups are not well integrated with each other (particularly the range management squadron and SLS). Option 2 would create a group- or wing-level integrator. Option 3 would create a maintenance group from maintenance resources in the operations group and launch control group, realign current launch control group leadership and/or staff positions, and realign all operations functions, including the current SLS, into the operations group.

The 45 RMS and SLS maintenance activities would be realigned into the newly formed maintenance group. Since 45 SCS does not perform or manage direct mission communications (unlike 30 SCS), it is not realigned. SLS maintenance functions would be reorganized into space maintenance units in a new SMXS, with each space maintenance unit providing support to its counterpart SLS. To provide common maintenance management functions for the maintenance group, a maintenance operations flight could be formed from resources in the new maintenance group, at the commander's discretion. The 21 MXG MOF provides an example of the functions and capabilities consolidated in a maintenance operations flight (plans and scheduling, training management, maintenance analysis, etc.). (See Appendix G for more details.)

Figure 5.5 shows the AS-IS organization of 45 SW. In 45 OG, the range management squadron provides maintenance oversight and support to the wing's range and launch system activities. The range

---

[7]   30 SW organizational options are not shown in the main report text because they are similar to those for 45 SW. See Appendix B for organizational options for HQ AFSPC, 21 SW, 30 SW, and 460 SW.

Table 5.3
Organizational Options for the 45th Space Wing

| Option 1: Baseline (AS-IS) | Option 2: Integrator at the | | Option 3: Form an MXG |
	Group Level (A)	Wing Level (B)	
Maintenance management is in 45 RMS and 45 LCG	Create a neutral supply-demand integrator on group staff Retain baseline structure in the operations and LCGs	Create a neutral supply-demand integrator on wing staff Retain baseline structure in the operations and LCGs	Rename LCG as MXG Pull maintenance functions out of SLS/LCSS and consolidate into a SMXS in the MXG Realign SLS squadrons to the operations group Move 45 RMS into MXG Create an MOF Reassign 45 SCS to 45 MSG

Figure 5.5
An AS-IS Graphical Representation of the 45th Space Wing

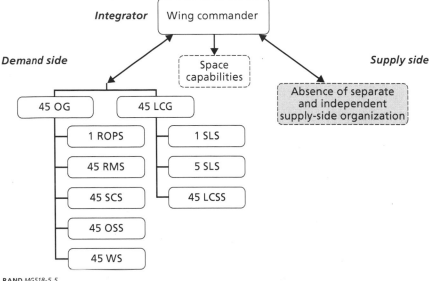

RAND *MG518-5.5*

flights providing QA, analysis, and contract oversight and management for the two major contracts—the Range Technical Services Contract (RTSC) and Launch Operations Support Contract. Appendix D briefly summarizes these contracts.

In 45 LCG, there are two primary mission SLSs and a launch support squadron (LCSS). Each SLS has a set of integrated teams for each platform (Delta II heritage or Delta IV/Atlas V Evolved Expendable Launch Vehicles); the teams include operations, maintenance, and acquisition personnel. As with the SLSs in 30 LCG, these squadrons are not organized into maintenance and operations sections. Figures 5.6 and 5.7 illustrate options 2A and 2B.

Figure 5.8 shows option 3 for 45 SW. This option would consolidate all wing-level maintenance functions into a new maintenance group that has been realigned using the launch control group overhead and that would move all operations (range and launch) functions into the operations group. Space maintenance units would be part of a new SMXS to provide maintenance support for the realigned SLSs. The

Figure 5.6
A Strategies-to-Tasks View of Option 2A for the 45th Space Wing

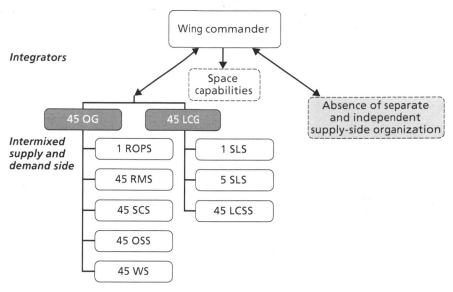

Integrators

Intermixed supply and demand side

RAND *MG518-5.6*

launch support squadron would become a launch support unit under the SMXS, supporting both SLSs. The SMXS would have a squadron commander, command section, and a maintenance officer and maintenance superintendent responsible for managing the production effort and standardizing processes across the space maintenance units and the launch support unit. The SMXS overhead positions could be obtained by making appropriate AFSC or grade changes of position, made possible by the savings accrued from consolidation of the realigning maintenance units (for example, applying the Air Staff 5-percent planning factor). This would allow the existing SLS staffs to transfer intact to the operations group. The group staff functions the launch support squadron currently performs would be realigned into the operations group's staff or operations support squadron in keeping with the CWO.

The range management squadron would move to the new maintenance group. Its maintenance functions (those associated with the Range Technical Service Contract and Launch Operations Support Contract), which directly support the range infrastructure and mission

Figure 5.7
A Strategies-to-Tasks View of Option 2B for the 45th Space Wing

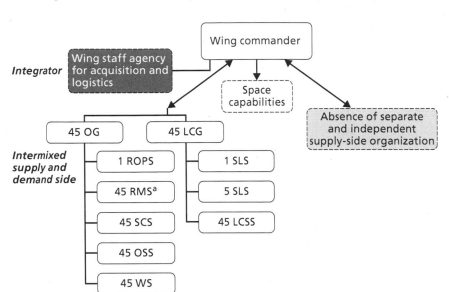

ª Realignment of O&M Management Flight (RMO) to new wing staff agency.
RAND MG518-5.7

equipment, would move to the maintenance group. The space communications squadron is reassigned to the mission support group to align with the CWO. A maintenance operations flight may be created to provide integrated planning and scheduling, training management, analysis, supply liaison, QA, etc., which are currently provided by functions dispersed throughout the operations group and launch control group.

Although the Joint Base Operations and Support Contract (J-BOSC) is managed jointly for Patrick AFB and NASA Cape Canaveral by the Cape Canaveral Spaceport Management Office (see Appendix D), the tasks directly affecting aircraft and spacecraft maintenance activities (for example, transient aircraft maintenance services) should be considered for breakout ("descoping") at the next contract review and/or renewal. These management tasks could be realigned to 45 RMS to give the maintenance group commander direct oversight of all

Figure 5.8
A Strategies-to-Tasks View of Option 3 for the 45th Space Wing

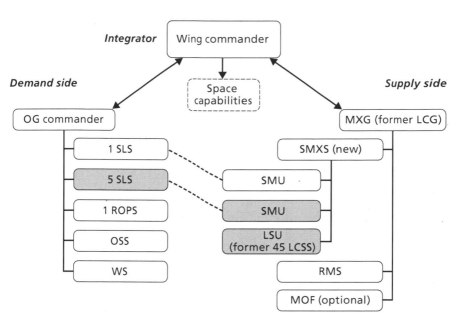

NOTE: MOF includes functions for Plans and Scheduling, Training Management, and Maintenance Analysis.
RAND *MG518-5.8*

maintenance activities. As an example, Appendix G provides a description of maintenance operations flight responsibilities.

Table 5.4 summarizes our analysis for 45 SW. We evaluated the three options using the same evaluation criteria we used for 50 SW.

We observed high levels of synergy and responsiveness to current operations with the baseline structure; however, the responsiveness was highly reactive. Option 2 is expected to sustain this and provide broader oversight at the wing level. Option 3 sustains high synergy and responsiveness, following a break-in period.

On fleet health, the team observed that the baseline structure commonly permits delays in scheduled maintenance and modifications because of the high value placed on operating limited range resources. During one interview session, we were told that a block of time in late calendar year 2005 had been required to be set aside for accomplish-

Table 5.4
Summary Evaluation of the 45th Space Wing Organizational Options

Evaluation Criteria	Option 1: Baseline (AS-IS)	Option 2: Integrator at the		Option 3: Form an MXG
		Group Level (A)	Wing Level (B)	
Operations-maintenance synergy and responsiveness to current operational requirements	High synergy, but reactive	High synergy Potential to be proactive	High synergy Potential to be proactive	High synergy Proactive
Fleet-health focus	No senior-level (O-6) advocacy Delays in scheduled maintenance and modifications common	Some improvement contingent on level of authority assigned to integrator	Some improvement contingent on level of authority assigned to integrator Has greater potential across the wing	Senior-level (O-6) advocacy Improved balance of long-term fleet health and current operations
Mentorship of space professionals	Limited	Limited	Limited	Improved by providing doctoral-level expert at MXG
Resources required	None additional	Integrator positions Some duplication necessary	Integrator positions	SMXS overhead positions,[a] facility space, and office equipment

[a] Sourced from manpower savings accrued through consolidation.

ing accumulated deferred maintenance. A broader, maintenance group O-6–level view of such consequences may offer alternatives that might not cause downtime and that may improve management and execution of sustainment activities.

Mentorship of Communications (33S) and Maintenance (21M) officers, particularly company-grade officers, is expected to improve if a colonel position (for example, a maintenance group commander) is created with direct authority for all wing-level maintenance.

Finally, implementing options 2 and 3 would require minimal resources. Option 2 would require realigning nine positions in the 45 RMS/RMO flight, one additional 2M0 enlisted position, and three additional operator positions (two 13S and one 1C6) for the integration organization. Since this option offers no consolidation savings, the additional positions would need to be sourced from elsewhere in the wing or command. Personnel assigned to the integration shop should be highly experienced. Civilians should fill most of the positions to provide continuity, which the changes of assignment of active duty personnel do not allow. These positions should be staffed with people with analytic skills and perhaps require, at minimum, a master of science in operational science from the Air Force Institute of Technology.

In keeping with the 45 SW UMDs and Air Force manpower standards for command staffs, option 3 would require realigning positions and potentially adding positions to create the SMXS overhead, depending on the savings accrued through consolidation. The Air Staff planning factor of 5-percent savings for consolidation efforts applied to the predicted size of the SMXS would fulfill the overhead requirements without creating an additional manpower bill. The existing launch control group command staff structure would transfer to the new maintenance group. A 109-person SMXS would be created from the maintenance positions in the current SLSs and launch support squadron. Its staff would be built from six positions recouped from consolidating all maintenance under one unit. Some grade and AFSC adjustments may be necessary. See Appendix C for the manpower data.

Implications

Adopting the strategies-to-tasks framework and philosophy can help guide the organizational development of AFSPC. Organizational changes at the headquarters have been made in accordance with the strategies-to-tasks framework, although these were not specifically so recognized at the time. The organization of the space wings could follow suit to improve the balance between current readiness and future readiness to support operations.

The AFSPC/A3 and each wing the operations group is responsible for carrying out such processes as the following:

- validating operational needs
- deconflicting competing space capabilities and determining priorities needed
- reviewing space options for meeting COCOM priorities.

AFSPC/A4S and each space wing maintenance group is responsible for the following:

- assessing infrastructure capabilities
- monitoring performance
- identifying system sustainment needs
- assessing efficiency and/or effectiveness of sustainment options.

Moving to the strategies-to-tasks framework and adopting its philosophy should enhance the quality of decisionmaking. Adopting the strategies-to-tasks philosophy recognizes the tension between the supply and demand organizations and the need for an integrator to analyze options. All this should help make the decisionmakers more informed. All the organizational structures—integrator at the group level, at the wing level, or a separate maintenance group—follow the expanded strategies-to-tasks framework and should enhance space-system sustainment. If the benefits of separating supply, demand, and integrator responsibilities are explicitly recognized, an organizational structure may not be as important.

CHAPTER SIX

Conclusions and Recommendations

As pointed out in this monograph, the strategies-to-tasks framework provides a conceptual foundation for developing a command philosophy for space-system support and, indeed, for supporting aircraft and missile systems as well. The strategies-to-tasks framework prescribes separating demand-side, supply-side, and integrator processes, remembering that these processes are often nested. The supply, demand, and integrator roles are defined not only at the execution level but at other levels as well—both inside and outside the command. Roles and responsibilities should be defined at all levels. From a command perspective, adopting this philosophy is an important first step. Adoption can provide a basis for enhancing processes, force development, doctrine, information systems, and organization in ways that can be sustained over time and through many leadership changes. The philosophy is timeless and should be set forth in doctrine. Figure 6.1 summarizes the options for improving support of space systems.

If the strategies-to-tasks framework and philosophy are adopted, implementing certain near-term actions should improve mission effectiveness without requiring substantial increases in resources. Indeed, most actions need leadership attention and do not add to costs significantly. The commander of AFSPC can indicate that he adopts the philosophy and that the headquarters has already taken actions to implement aspects of it. The commander can then direct each space wing to organize according to the strategies-to-tasks framework and to align its processes accordingly. All the organizational structures—integrator at the group level, integrator at the wing level, or a separate maintenance group—

Figure 6.1
Analysis Suggests Options for Improving Support of Space Systems

- **Recognize** philosophy → **Explicitly** identify integrator, demand, and supply roles **at HQ AFSPC, wings, SMC/ESC**

- **Expand** doctrine → Delineate roles of operations, maintenance, communications, and plans in AFDD 2-2, AFDD 2-4, and TTPs

- **Enhance** processes → Focus **sustainment planning and execution processes** on outcomes and **identify and** separate supply, demand, and integrator processes

- **Refine** force development → Encourage space support professional development **and multiple tours in AFSPC**

- **Develop** tools and systems → Identify and define metrics and information **needed to manage and relate sustainment actions to operational capabilities**

- **Strengthen** organizations respectively → Assign supply, demand, and integrator processes to operations, maintenance, and plans organizations **and staff with trained professionals**

RAND *MG518-6.1*

should follow the expanded strategies-to-tasks framework and should enhance space-system sustainment. However, if the strategies-to-tasks benefits are explicitly recognized, an organizational structure may not be as important.

The commander can also attend monthly meetings on space-system sustainment and ensure that leading sustainment indicators are being used and that operations and sustainment issues are both being addressed. He can also encourage force-development efforts aimed at implementing the strategies-to-tasks framework and enhancing management of space-system sustainment. In the longer term, metrics, the POM process, and information systems can be enhanced along the lines suggested in this monograph.

To help support space-system sustainment funding requirements, all decisionmakers need a better understanding of how space capabilities affect military joint operational effects and contribute to the public good. In the POM process, common-use systems, such as GPS, and space contributions to the warfighter, such as space surveillance and weather, need better representation. These services and contributions, provided by the Air Force, are not well understood, yet they are important to many users—military and civilian. These contributions need to be explained and understood in the POM process as sustainment funds are requested.

The Strategies-to-Tasks Framework

History

RAND developed the strategies-to-tasks framework in the late 1980s (see Kent, 1989, and Thaler, 1993), and DoD has widely applied it to aid in strategy development, campaign analysis, and modernization planning.[1] The framework has proven useful for providing intellectual structure to ill-defined or complex problems. If used correctly, it links resources to specific military tasks that require resources, which in turn are linked hierarchically to higher-level operational and national-security objectives (see Figure A.1). Working through the strategies-to-tasks hierarchy can help identify areas needing new capabilities, clarify responsibilities among actors contributing to accomplishing a task or an objective, and place the contributions of multiple entities and organizations working to achieve some common objective into a common framework.

Strategies-to-Tasks Hierarchies

At the highest levels of the strategies-to-tasks hierarchy, we consider *national goals*, which are derived from U.S. heritage and are embodied in the U.S. Constitution (see Figure A.1). These national goals do not

---

[1] Internal examples are Lewis, Coggin, and Roll (1994) and Niblack, Szayna, and Bordeaux (1996). Outside RAND, the framework is in use by the Air Force, the Army, and elements of the Joint Staff.

Figure A.1
Strategies-to-Tasks Hierarchy

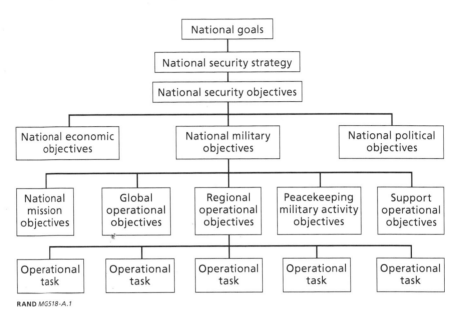

RAND *MG518-A.1*

change over time. They form the foundation from which all U.S. statements on national security are derived.

National security strategy is formulated in the executive branch. It outlines strategy for applying the national instruments of power—political, economic, military, and diplomatic—to achieve U.S. national security objectives.

National security objectives define what must be done to preserve and protect our fundamental goals and interests from threats and challenges that originate abroad. In contrast to national goals, national security objectives change in accordance with changes in the geopolitical environment.

National military objectives are formulated by the Secretary of Defense and the Chairman of the Joint Chiefs of Staff. The national military objectives define how the United States will apply military power to attain the national objectives that support the national security strategy. Collectively, they define the national military strategy, which identifies (at a high level) how the United States will respond to threats to its national security.

Operational objectives describe how forces will be used to support the national military objectives and may be regional or global and include support activities necessary to sustain military operations.

Tasks, formulated by the COCOMs and their staffs, are the specific functions that must be performed to accomplish an operational objective. Operational tasks constitute the building blocks of the application of military power. Examples of tasks that might be accomplished to help achieve the operational objective of suppressing the generation of enemy air sorties include the following (from Kent, 1989, and Thaler, 1993):

- crater and/or mine runways and taxiways
- destroy aircraft in the open or in revetments
- destroy key hardened support facilities
- destroy aircraft in hardened shelters.

In this analysis, we use the strategies-to-tasks framework to show how space-system sustainment can be related to the task-organized operational elements used to accomplish operational tasks (or, in other words, to create desired joint operational effects). This strategies-to-tasks construct outlines how national goals can be disaggregated into national diplomatic, economic, informational, and military objectives. Regional military operational objectives can be formulated from national military objectives. Joint operational tasks can be assigned to joint task forces in the region.[2] Task-organized operational elements carry out the tasks assigned to them during different periods, and task-organized combat support elements provide the needed support to conduct the operational mission. In this framework, space-system sustainment would be one of the task-organized combat support elements needed to create the operational effects. Space-system elements could be called upon to provide weather data or troop positioning data or to

---

[2] The number and nature of these joint operational tasks will change over time.

provide intelligence data needed to sustained operations in the field and, thereby, contribute to the accomplishment of operational tasks.

Sustainment Organization Structure Options for HQ AFSPC and the 21st, 30th, and 460th Space Wings

This appendix presents organizational options for headquarters AFSPC and for the space wings not described in the body of the monograph: the 21, 30, and 460 SWs.

HQ AFSPC Organization

AFSPC is headquartered at Peterson AFB, Colorado. Tables B.1 and B.2 detail the options for this unit and our evaluation of them. Figures B.1 through B.3 illustrate the baseline (AS-IS) organization and those for options 2 and 3.

Table B.1
Organizational Options for HQ AFSPC

Option 1: Baseline (AS-IS)	Option 2: Create Permanent A4S Positions in A3	Option 3: Realign Sustainment Responsibilities from A3 into A4S
Operations (A3) and Maintenance (A4A6) not tightly integrated with respect to sustainment decisions and priorities	Realign A4S expertise in A3 divisions to improve integration	Realign responsibilities for sustainment planning, programming, and funding out of A3 into A4S Strengthen integration role of A8A9 (XP)

Figure B.1
An AS-IS Graphical Representation of HQ AFSPC

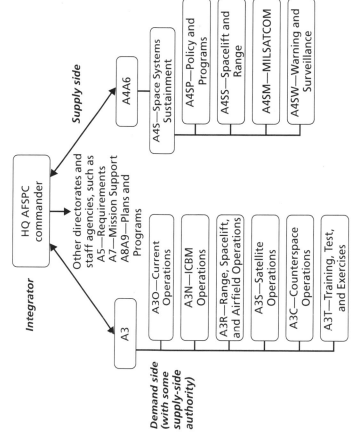

Integrator

HQ AFSPC commander

Supply side

A4A6

A4S—Space Systems Sustainment

A4SP—Policy and Programs

A4SS—Spacelift and Range

A4SM—MILSATCOM

A4SW—Warning and Surveillance

Other directorates and staff agencies, such as
A5—Requirements
A7—Mission Support
A8A9—Plans and Programs

Demand side (with some supply-side authority)

A3

A3O—Current Operations

A3N—ICBM Operations

A3R—Range, Spacelift, and Airfield Operations

A3S—Satellite Operations

A3C—Counterspace Operations

A3T—Training, Test, and Exercises

RAND *MG518-B.1*

Figure B.2
A Strategies-to-Tasks View of Option 2 for HQ AFSPC

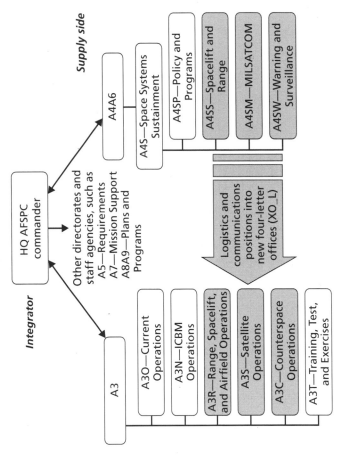

RAND *MG518-B.2*

Figure B.3
A Strategies-to-Tasks View of Option 3 for HQ AFSPC

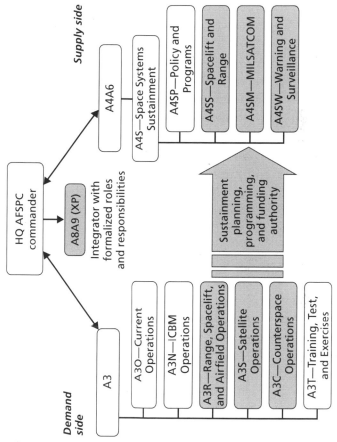

RAND *MG518-B.3*

Table B.2
Summary Evaluation of HQ AFSPC Organizational Options

Evaluation Criteria	Option 1: Baseline (AS-IS)	Option 2: Create Permanent A4S Positions in A3	Option 3: Realign sustainment responsibilities from A3 to A4S
Effectiveness of POMing for sustainment	Disjointed within and across systems[a]	Improved focus within systems	Improved focus within and across systems
Prevention of "sustainment surprises"	Marginal	Potential to improve	Potential to improve
Resources required	None additional	Realign positions from A4S to A3	Consider realigning PEM/budgeteer positions (if any) from A3 to A4S

[a] Separated from logistics and communications functional staff expertise.

21st Space Wing Organization

The 21 SW provides missile warning and space control to North American Aerospace Defense Command and U.S. Strategic Command using Defense Support Program (DSP) satellites. Tables B.3 and B.4 detail the options for this unit and our evaluation of them. Figures B.4 through B.6 illustrate the baseline (AS-IS) organization and those for options 2A, 2B, and 3.

Table B.3
Organizational Options for the 21st Space Wing

Option 1: Baseline (AS-IS)	Option 2: Integrator at the		Option 3: Expand MXG
	Group Level (A)	Wing Level (B)	
Maintenance activities exist in each SWS/SPCS in the OG	Create a neutral supply-demand integrator on OG/MXG	Create a neutral supply-demand integrator on wing staff	Pull maintenance functions out of SWS/SPCS and consolidate into an SMXS in the MXG
Some consolidated maintenance functions exist in the MXG	Retain baseline structure in OG and MXG	Retain baseline structure in OG and MXG	Expand PMD into a space management squadron[a]

[a] Similar to the range management squadron in range space wings.

Figure B.4
An AS-IS Graphical Representation of the 21st Space Wing

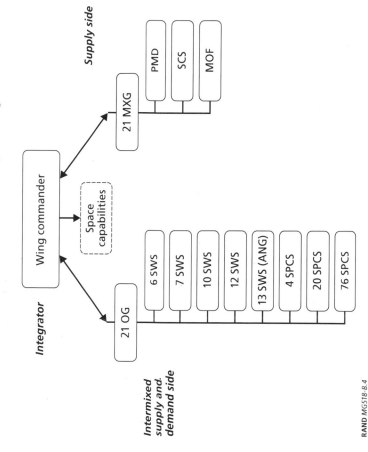

RAND MG518-B.4

Figure B.5
A Strategies-to-Tasks View of Option 2A for the 21st Space Wing

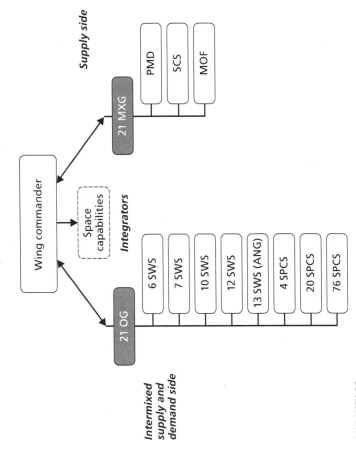

Figure B.6
A Strategies-to-Tasks View of Option 2B for the 21st Space Wing

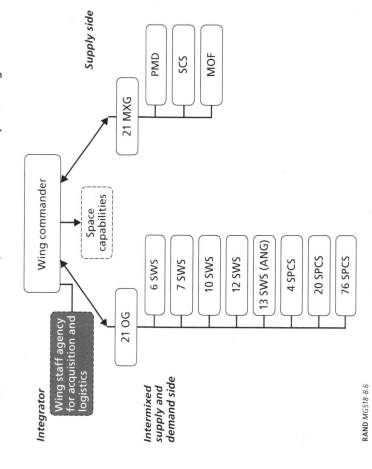

Figure B.7
A Strategies-to-Tasks View of Option 3 for the 21st Space Wing

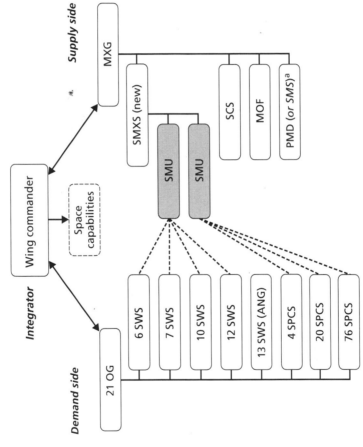

NOTES: Because units are geographically separated, separate SMXSs may not be feasible for space warning and space control. SMS is parallel to 30/45 RMS.

RAND *MG518-B.7*

Table B.4
Summary Evaluation of the 21st Space Wing Organizational Options

Evaluation Criteria	Option 1: Baseline (AS-IS)	Option 2: Integrator at the		Option 3: Expand MXG
		Group Level (A)	Wing Level (B)	
Operations-maintenance synergy and responsiveness to current operational requirements	High synergy, but reactive	High synergy Potential to be proactive	High synergy Potential to be proactive	High synergy, proactive
Fleet-health focus	Senior-level advocacy hindered by lines of authority (operations as opposed to maintenance group) Delays in scheduled maintenance and modifications occur	Some improvement contingent on level of authority assigned to integrator	Some improvement contingent on level of authority assigned to integrator Has greater potential across the wing	Senior-level (O-6) advocacy with authority Improved balance of long-term fleet health and current operations
Mentorship of space supply-side communications and maintenance personnel	Adequate for personnel in MXG	See above	See above	Improved mentorship for realigned functions (now under doctoral-level in supply side)
Resources required	None additional	Integrator positions Some duplication is necessary	Integrator positions	SMXS overhead positions,[a] integrator position, office space, office equipment

[a] Sourced from manpower savings accrued through consolidation.

30th Space Wing Organization

This wing provides command and control of the Western Range capabilities. Note that, if the mission support group is responsible for maintaining portions of the range and launch infrastructure, it should be included in this organizational structure. Applying the strategies-to-tasks framework, integration could occur at the mission support group commander level or at the wing level, or personnel involved in range or launch sustainment could be rolled into a new maintenance group.

Tables B.5 and B.6 detail the options for this unit and our evaluation of them. Figures B.8 through B.11 illustrate the baseline (AS-IS) organization and those for options 2A, 2B, and 3.

Table B.5
Organizational Options for the 30th Space Wing

| Option 1:
Baseline (AS-IS) | Option 2:
Integrator at the | | Option 3:
Form an MXG |
	Group Level (A)	Wing Level (B)	
Maintenance management resides in 30 RMS and 30 SCS under the OG and in the LCG	Create a neutral supply-demand integrator on group staff	Create a neutral supply-demand integrator on wing staff	Rename LCG as MXG; pull maintenance functions out of SLS and consolidate into an SMXS in the MXG
	Retain baseline structure in the OG and LCG	Retain baseline structure in the OG and LCG	Realign SLS squadrons to the OG
			Move 30 RMS and 30 SCS into MXG
			Return non–direct-mission communications resources from the SCS to 30 MSG
			Create an MOF

Figure B.8
An AS-IS Graphical Representation of the 30th Space Wing

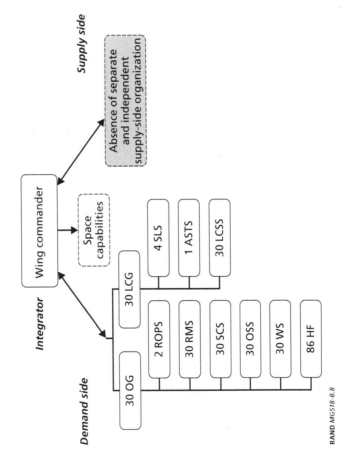

RAND MG518-B.8

Figure B.9
A Strategies-to-Tasks View of Option 2A for the 30th Space Wing

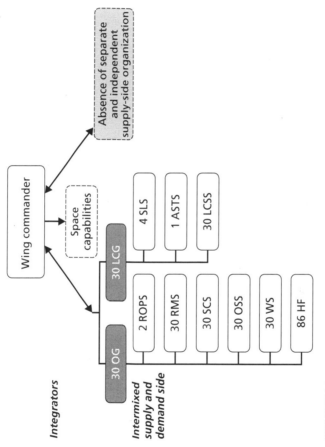

RAND *MG518-B.9*

Figure B.10
A Strategies-to-Tasks View of Option 2B for the 30th Space Wing

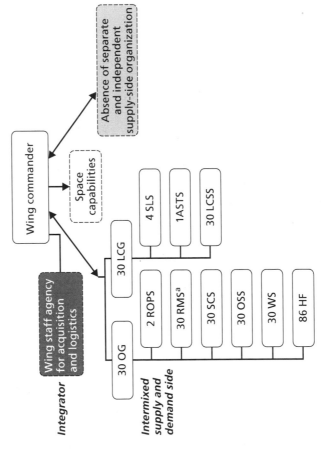

Wing commander

Absence of separate and independent supply-side organization

Space capabilities

Integrator — Wing staff agency for acquisition and logistics

30 OG

30 LCG

Intermixed supply and demand side

2 ROPS
30 RMS[a]
30 SCS
30 OSS
30 WS
86 HF

4 SLS
1ASTS
30 LCSS

[a] Realignment of O&M Management Flight (RMO) to new wing staff agency.

RAND *MG518-B.10*

Figure B.11
A Strategies-to-Tasks View of Option 3 for the 30th Space Wing

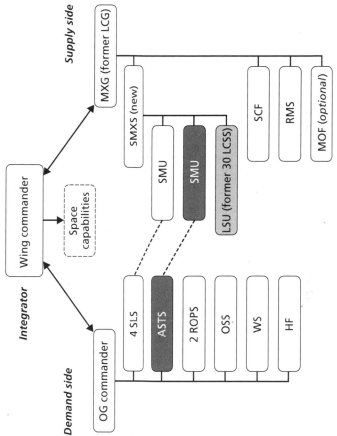

NOTE: MOF includes functions for Plans and Scheduling, Training Management, and Maintenance Analysis.

RAND *MG518-B.11*

Table B.6
Summary Evaluation of the 30th Space Wing Organizational Options

Evaluation Criteria	Option 1: Baseline (AS-IS)	Option 2: Integrator		Option 3: Form an MXG
		At the Group Level (A)	At the Wing Level (B)	
Operations-maintenance synergy and responsiveness to current operational requirements	High synergy, but reactive	High synergy Potential to be proactive	High synergy Potential to be proactive	High synergy, proactive
Fleet-health focus	No senior-level (O-6) advocacy Delays in scheduled maintenance and modifications common	Some improvement contingent on level of authority assigned to integrator	Some improvement contingent on level of authority assigned to integrator Has greater potential across the wing	Senior-level (O-6) advocacy Improved balance of long-term fleet health and current operations
Mentorship of space communications and missile maintenance officers	Limited	Limited	Limited	Improved by providing doctoral-level expert at MXG
Resources required	None additional	Integrator positions Some duplication is necessary	Integrator positions	SMXS overhead positions,[a] facility space, and office equipment

[a] Sourced from manpower savings accrued through consolidation.

460th Space Wing Organization

The 460 SW provides global surveillance and missile warning by operating the Defense Support Program satellite.

Note that, since 460 SW has only one active-duty operations squadron, creation of an SMXS is not warranted for option 3. Should additional operations squadrons be activated, additional SMUs could be added and an SMXS structure created as an intermediate command echelon between SMU and MXG (as proposed for the 21 SW and 50 SW). Alternatively, if a Future Total Force approach is taken, an SMXS could be constructed with active-duty, AFRC, and ANG SMUs, respectively, to provide support to the 2, 8, and 137 SWSs.

Tables B.7 and B.8 detail the options for this unit and our evaluation of them. Figures B.12 through B.15 illustrate the baseline (AS-IS) organization and those for options 2A, 2B, and 3.

Table B.7
Organizational Options for the 460th Space Wing

Option 1: Baseline (AS-IS)	Option 2: Integrator at the		Option 3: Create an MXG
	Group Level (A)	Wing Level (B)	
Maintenance activities conducted by the system program office	Create a neutral supply-demand integrator on group staff	Create a neutral supply-demand integrator on wing staff	Create an MXG
The SCS is under the OG	Retain baseline structure in the OG	Retain baseline structure in the OG	Establish a SMU in the MGX to work with SMC on maintenance issues
			Extract direct mission communications functions from the SCS and realign into a space communications flight (SCF) in the MXG
			Realign the remainder of the SCS to the MSG
			Create an MOF in the MXG

Figure B.12
An AS-IS Graphical Representation of the 460th Space Wing

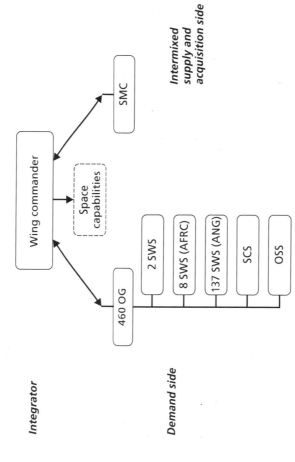

Integrator

Wing commander

SMC

Space capabilities

Intermixed supply and acquisition side

Demand side

460 OG

2 SWS

8 SWS (AFRC)

137 SWS (ANG)

SCS

OSS

RAND *MG518-B.12*

Figure B.13
A Strategies-to-Tasks View of Option 2A for the 460th Space Wing

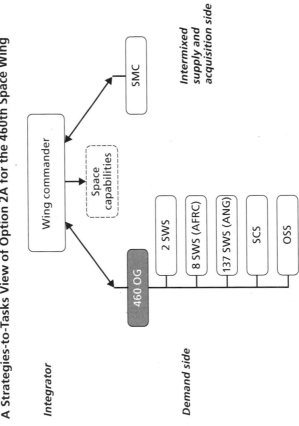

Integrator

Wing commander

SMC

Intermixed supply and acquisition side

Space capabilities

460 OG

2 SWS

8 SWS (AFRC)

137 SWS (ANG)

SCS

OSS

Demand side

RAND *MG518-B.13*

Figure B.14
A Strategies-to-Tasks View of Option 2B for the 460th Space Wing

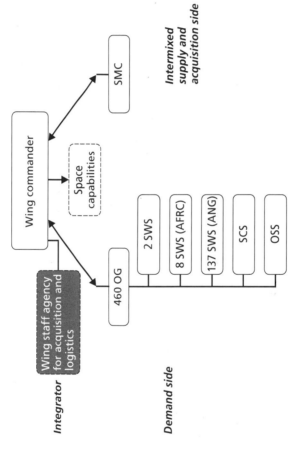

Figure B.15
A Strategies-to-Tasks View of Option 3 for the 460th Space Wing

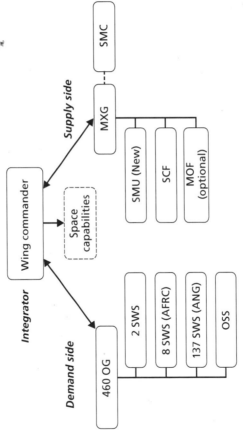

NOTE: MOF includes functions for Plans and Scheduling, Training Management, and Maintenance Analysis.

RAND MG518-B.15

Table B.8
Summary Evaluation of the 460th Space Wing Organizational Options

Evaluation Criteria	Option 1: Baseline (AS-IS)	Option 2: Integrator at the		Option 3: Form an MXG
		Group Level (A)	Wing Level (B)	
Operations-maintenance synergy and responsiveness to current operational requirements	High synergy, but reactive	High synergy Potential to be proactive	High synergy Potential to be proactive	High synergy, proactive
Fleet-health focus	No independent senior-level advocacy Delays in scheduled maintenance and modifications occur	Some improvement contingent on level of authority assigned to integrator	Some improvement contingent on level of authority assigned to integrator	Senior-level (O-6) advocacy Improved balance of long-term fleet health and current operations
Mentorship of space supply-side communications and maintenance personnel	Limited	Limited	Limited	Improved by providing doctoral-level expert at MXG
Resources required	None additional	Integrator positions	Integrator positions	MXG staff positions, a office space, office equipment

a Estimate six to eight positions required based on comparable small MXGs.

Manpower Analysis

This appendix identifies manpower requirements for the space wing organizational options presented in Chapter Five. Manpower data are derived from fiscal year 2005 UMDs provided by HQ AFSPC and tabulated for the two- or three-digit AFSC.[1]

Tables C.1 through C.3 examine options for 21 SW, Tables C.4 through C.7 for 30 SW; Tables C.8 through C.10 for 45 SW, Tables C.11 through C.15 for 50 SW, and Tables C.16 through C.20 for 460 SW.

Finally, Table C.21 compares the options for all these wings.

---

[1] UMD data were baselined using funded fiscal year 2005 positions.

Organizational Options for the 21st Space Wing

Table C.1
21st Space Wing Organizational Option 2:
Creation of Integration Office on Wing Staff

Air Force Specialty	Realigned	Source
Space and Missile Operations (13S)	2	21 OG
Space Systems Operations (1C6)	1	21 OG
Munitions and Missile Maintenance (21M)	1	21 OG
Communications (3C or 2E)	1	21 OG
Communications and Information (33S)	1	21 OG
Scientist and Operations Analysis (61)	1	21 OG
Total	7	

Table C.2
21st Space Wing
Organizational Option 3:
Expand the Maintenance Group

Unit	Realigned
SMXS	301
Total	301

NOTE: See Table C.3

Table C.3
Creation of a Space Maintenance Squadron

Air Force Specialty	Realigned	Sources
Aerospace Ground Equipment (2A)	14	4 SPCS, 76 SPCS
Satellite and Wideband Communications (2E)	147	6, 7, 10, 12 SWS, 4, 20, 76 SPCS
Supply (2S0)	15	6, 7, 10 SWS, 4, 20, 76 SPCS
Communications and Computer Systems (3C)	78	6, 7, 10, 12 SWS, 4, 20, 76 SPCS
Other Maintenance and/or Logistics	4	6 SWS, 4, 20, 76 SPCS
Munitions and Missile Maintenance (21M)	3	4, 76 SPCS
Communications and Information (33S)	28	6, 7, 10, 12 SWS, 20 SPCS
Scientist and Operations Analyst (61)	3	4, 76 SPCS
Information Manager (3A0)	9	6, 7, 10, 12 SWS, 4, 20, 76 SPCS
Total	301	

Organizational Options for the 30th Space Wing

Note that creation of an MOF (Table C.5) allows realignment and consolidation of such functions as plans and scheduling, training management, maintenance analysis, supply liaison, and staff support. Manpower resources for an MOF would be drawn from the other squadrons and flights in the group and therefore would not affect the total manpower.

Table C.4
30th Space Wing Organizational Option 2:
Creation of an Integration Office on Wing Staff

Air Force Specialty	Realigned	Source
Space and Missile Operations (13S)	2	30 OG
Space System Operations (1C6)	1	30 OG
Munitions and Missile Maintenance (21M)	1	30 LCG
Missile and Space Sys Maintenance (2M0)	1	30 LCG
Communications and Information (33S)	1	30 RMS
Acquisition Manager (63A)	25	30 RMS
Total	30	

Table C.5
30th Space Wing Organizational Option 3:
Creation of a Maintenance Group

Unit	Realigned	Notes
MXG staff	8	Realign from 30 LCG staff.
SMXS	125	See Table C.6.
RMS	78	Unit realignment.
SCF	65	See Table C.7.
MOF	TBD	
Total	276	

Table C.6
Creation of a Space Maintenance Squadron

Air Force Specialty	Realigned	Sources
Missile and Space System Maintenance (2M0)	64	30 LCSS,[a] 4 SLS, 1 ASTS
Munitions and Missile Maintenance (21M)	3	30 LCSS
Engineer (62E)	42	30 LCSS, 4 SLS, 1 ASTS
Acquisition Manager (63A)	11	30 LCSS, 4 SLS, 1 ASTS
Info Management (3A0)	5	30 LCSS, 4 SLS
Total	125	

[a] As of November 1, 2005, 2 SLS was disbanded and replaced by 30 LCSS.

Table C.7
Creation of a Space Communications Flight

Air Force Specialty	Realigned	Source
Metals Technology (2A7)	1	30 SCS
Communications (2E and 3C)	53	30 SCS
Supply (2S0)	1	30 SCS
Munitions and Missile Maintenance (21M)	1	30 SCS
Logistics Readiness (21R)	1	30 SCS
Communications and Information (33S)	6	30 SCS
Engineer (62E)	2	30 SCS
Total	65	

Organizational Options for the 45th Space Wing

The majority of realigned positions for the 45 SW are sourced from the Range Management Operations flight of the range management squadron. The differences between option 2 for 30 SW and 45 SW derive from the scope of contracts administered; for example, the infrastructure contract at Vandenberg AFB is managed by the wing, but the one at Patrick AFB is managed externally by a joint organization that also supports NASA (see Appendix B).

Note that the creation of an MOF (Table C.9) allows realignment and consolidation of such functions as plans and scheduling, training management, maintenance analysis, supply liaison, and staff support. Manpower resources for an MOF would be drawn from the other squadrons and flights in the group and therefore would not affect the total manpower.

Table C.8
45th Space Wing Organizational Option 2:
Creation of an Integration Office on Wing Staff

Air Force Specialty	Realigned	Source
Space and Missile Operations (13S)	2	45 OG
Space System Operations (1C6)	1	45 OG
Munitions and Missile Maintenance (21M)	1	45 RMS
Missile and Space System Maintenance (2M0)	1	45 LCG
Communications and Information (33S)	4	45 RMS
Engineer (62E)	4	45 RMS
Total	13	

**Table C.9
45th Space Wing Organizational Option 3:
Creation of a Maintenance Group**

Unit	Realigned	Notes
MXG staff	8	Realign from 45 LCG staff.
SMXS	109	See Table C.10.
RMS	118	Unit realignment.
MOF	TBD	
Total	235	

**Table C.10
Creation of a Space Maintenance Squadron**

Air Force Specialty	Realigned	Source
Missile and Space System Maintenance (2M0)	65	1 SLS, 5 SLS, 45 LCSS
Munitions and Missile Maintenance (21M)	4	1 SLS, 45 LCSS
Engineer (62E)	31	1 SLS, 5 SLS, 45 LCSS
Acquisition Manager (63A)	9	5 SLS, 45 LCSS
Total	109	

Organizational Options for the 50th Space Wing

Note that the space communications squadron merger in February 2006 will have minimal effect if the recommendation to realign non–direct-mission support to the mission support group is followed.

Also note that, if the entire 50 SCS were retained in a new MXG (as in 21 SW), the proposed MXG would be 81 percent larger, to 401 authorized positions:

850 SCS authorized = 86
850 SCS command staff = 4
50 SCS authorized = 192
50 SCS command staff = 7
Original 50 SCS to transfer = 13
Merged command staff authorized = 5
Command staff adjustment = −6
Merged SCS authorized = 86 − 4 + 192 − 7 + 5 = 272
Original proposed authorized = 92 (including attachment MOF)
Delta = +180
Original proposed MXG authorized = 221
Delta percentage = (221 + 180)/221 − 1 = 81 percent larger.

Table C.11
50th Space Wing Organizational Option 2:
Creation of an Integration Office on Wing Staff

Air Force Specialty	Realigned	Sources
Space and Missile Operations (13S)	1	50 OG or 50 NOG
Space System Operations (1C6)	1	50 OG or 50 NOG
Communications (2E1)	1	850 SCS
Communications and Information (33S)	1	850 SCS
Engineer (62E)	1	50 NOG
Total	5	

Table C.12
50th Space Wing Organizational Option 3:
Creation of a Maintenance Group

Unit	Realigned	Notes
MXG staff	30	Realigned from 50 NOG staff.
SMXS	99	See Table C.13.
SCS	76	See Table C.14.
MOF	16	See Table C.15.
Total	221	

Table C.13
Creation of a Space Maintenance Squadron

Air Force Specialty	Realigned	Sources
Communications (2E and 3C)	46	1, 2, 3, 4, 21, and 22 SOPS
Communications and Information (33S)	46	1, 2, 3, 4, and 21 SOPS
Engineer (62E)	2	2 and 21 SOPS
Info Management (3A0)	5	1, 2, 3, 4, and 21 SOPS
Total	99	

Table C.14
Realignment of a Space Communications Squadron (850 SCS)

Air Force Specialty	Realigned	Sources
Communications (2E and 3C)	38	50 and 850 SCS
Communications and Information (33S)	27	50 and 850 SCS
Engineer (62E)	6	50 and 850 SCS
Info Management (3A0)	5	50 and 850 SCS
Total	76	

Table C.15
Creation of a Maintenance Operations Flight

Air Force Specialty	Realigned	Source
Maintenance Management Analyst (2R)	5	850 SCS
Supply (2S0)	4	850 SCS
Munitions and Missile Maintenance (21M)	1	850 SCS
Logistics Readiness (21R)	6	850 SCS
Total	16	

NOTE: Additional functions and capabilities in the MXG may also be reassigned to the MOF as desired within existing resource constraints.

Organizational Options for the 460th Space Wing

Note that the total manpower for the proposed MXG (Table C.17) is comparatively low (less than 100 positions). One adaptation to preserve the strategies-to-tasks nature would be to create an SMSX integrating the SMU, SCF, and optional MOF. See comments for Table C.19 on other considerations.

Also note that, as proposed, the separation of direct mission-support communications maintenance from base operating support (e.g., local area network, telephones, audiovisual support) follows the construct proposed for 30 SW option 3 (Table C.20). As an alternative, to provide a more-robust MXG, the entire 339-person SCS could be realigned as a whole. The latter would align with 21 MXG AS-IS and 50 SW option 3 constructs.

Table C.16
460th Space Wing Organizational Option 2:
Creation of an Integration Office on Wing Staff

Air Force Specialty	Realigned	Source
Space and Missile Operations (13S)	1	460 OG
Space System Operations (1C6)	1	460 OG
Communications (3C)	1	460 OG
Communications and Information (33S)	1	460 OG
Total	4	

Table C.17
460th Space Wing Organizational Option 3:
Creation of a Maintenance Group

Unit	Realigned	Notes
XMG overhead	6	See Table C.18.
SMU	22	See Table C.19.
SCF	29	See Table C.20.
Total	57	

Table C.18
Creation of a Maintenance Group

Air Force Specialty	Realigned	Source
Logistics Commander (20C0)	1	See note
Executive Officer (33S)	1	See note
Superintendent (9G1)	1	See note
Info Manager (3A0)	2	See note
Other staff (optional)	1	See note
Total	6	

NOTE: MXG command section positions cannot reasonably be sourced through consolidation or realignment. AFSPC or 460 SW would have to provide the positions necessary to create the group overhead structure.

Table C.19
Creation of a Space Maintenance Unit

Air Force Specialty	Realigned	Source
Satellite and Wideband Communications (2E)	1	2 SWS
Communications and Computer Systems (3C)	13	2 SWS
Communications and Information (33S)	6	2 SWS
Info Manager (3A0)	2	2 SWS
Total	22	

Table C.20
Creation of a Space Communications Flight

Air Force Specialty	Realigned	Source
Communications (2E)	29	460 SCS
Total	29	

Comparing the Wings

Table C.21
Comparison of Space Wing Organizational Options

Organizational Option	21 SW	30 SW	45 SW	50 SW	460 SW
Option 2: Integration Office[a]	7	30	13	5	4
Option 3: Creation of MXG					
MXG staff[a]	N/A	8	8	30	6
SMXS	301	125	109	99	22
RMS	N/A	78	118	N/A	N/A
SCS or SCF	N/A	65	N/A	76	29
MOF	N/A	TBD	TBD	16	N/A
Total	301	276	235	221	57

[a] The large differences between group staff sizes are due to the 16 Acquisition Manager (63A) and four Logistics Readiness Officer (21R) positions on the 50 NOG staff. The functions associated with these positions could be realigned to the SCS or MOF, as appropriate.

Summary of Air Force Space Command Range Service-Support Contracts

This appendix briefly describes the major range service-support contracts at Vandenberg AFB and at the Kennedy Spaceport and Patrick AFB. The Spacelift Range System Contract (SLRSC), which is managed by SMC/RN and provides sustaining engineering, modernization, and level-2 maintenance for the Launch and Test Range Architecture of the ranges and remote sites, is not discussed here.

30th Space Wing, Vandenberg AFB, California

Four major contracts support 30 SW O&M range and space launch activities. Each contract is managed by 30 RMS.

- Western Range Operations Communications and Information
  - provides O&M of range facilities and equipment, including control centers, radar systems, telemetry systems, optical systems, missile flight-termination systems, surveillance systems, and weather systems
  - supports planning and scheduling, flight safety, technical studies, and postlaunch performance analysis
  - provides preventive and/or corrective maintenance of range systems and facilities
  - provides services for networks, ground radio, telephone, voice and data switching, antennas, and long-haul communications

- provides O&M support for space launch complexes, control centers, and test facilities
- provides electronic security and physical security support at various facilities
- provides remote base operations.
- Launch Operations Support Contract
  - provides mission infrastructure support for the Launch and Test Range System, including maintenance and modification of nonprime mission equipment, launch support facilities, and aerospace ground equipment.
- Aerospace Support Services Contract
  - provides O&M services for the Aerospace Maintenance Operations Center, unconventional propellant handling and storage, protective equipment, Precision Measurement Equipment Laboratory (Type IIC), transient aircraft maintenance, flightline aerospace ground equipment, and fleet management for all vehicles leased or provided as government-furnished equipment.
- Wing Safety Performance Evaluation Contract
  - develops and uses flight safety analysis tools, ensures compliance with DoD directives and conformance to range safety requirements and industry standards
  - oversees space and missile program ground activities, including evaluation of hazardous procedures, such as vehicle and payload processing and recovery
  - supports facility and infrastructure safety evaluations
  - assesses wing operational systems
  - maintains a technical library.

45th Space Wing, Patrick AFB, Florida

Three major contracts support 45 SW O&M range and space launch activities. Two of the contracts are managed by 45 RMS, while the third (J-BOSC) is managed by a joint Cape Canaveral Spaceport

Management Office reporting to both the 45 SW commander and the Director of the Kennedy Space Center.

- Range Technical Services Contract (RTSC)
  - provides O&M of technical spaceflight tracking and control facilities and equipment, including spacelift range instrumentation systems, such as radar, optics, and telemetry; communication systems for secure data, voice, video, satellite communication, and microwaves; radio at high through ultrahigh high frequencies; and enterprise information technology systems
  - supports mission planning and analysis, range operations scheduling, financial systems and requirements analysis, range safety, and remote base operations
  - This contract is being replaced by the Eastern Range Technical Services contract, which has a similar scope.
- Launch Operations Support Contract
  - provides O&M of critical facilities and systems, systems management, safety engineering, ordnance handling and storage, spaceport utilization scheduling, spaceport information systems, operational training and badging, and visitor-center services.
- Joint Base Operations and Support Contract (J-BOSC)
  - provides infrastructure O&M, supply support, vehicle maintenance, transient aircraft maintenance (including the Spaceport and Patrick AFB), Precision Measurement Equipment Laboratory, propellant services, life support, airfield services, roads and groundskeeping, engineering and technical services, firefighting, security, medical and environmental services, and refuse collection.

Table D.1
Comparison of Major Contracts

Function	30 SW	45 SW
O&M of critical mission equipment and infrastructure	Western Range Operations Communications and Information	RTSC[a]
O&M of noncritical mission equipment and infrastructure	Launch Operations Support Contract	Launch Operations Support Contract
Base operating support	Aerospace Support Services Contract	J-BOSC
Wing safety (flight and ground)	Wing Safety Performance Evaluation Contract	N/A[b]

[a] Transitioning to Eastern Range Technical Services.

[b] Although safety is a performance requirement of each contract, this function is provided by other independent contracts managed by the Wing Safety office.

Comparison of Air Combat Command and Air Force Space Command Range Service-Support Contracts

This appendix briefly describes the processes Air Combat Command (ACC) uses to manage its Primary Training Ranges (PTRs) and compares them to those for AFSPC's Eastern and Western Ranges.

The ACC PTRs are operated and maintained under a single firm-fixed-price umbrella contract administered by the ACC Contracting Squadron and with the Ranges, Airspace, and Airfield Operations Division (ACC/A3A) providing program management. The contract provides O&M support for electronic equipment, targets, grounds, and facilities. Operational control of range systems is the responsibility of each range operating agency (e.g., range wing). Range metrics for system performance and contract performance assist in managing the contractual effort.[1]

Range wings are not wholly patterned after the CWO structure. For example, there is no maintenance group, but there are a mission support group and an operations group. In the range wing structure, a range squadron under the operations group oversees and is the single focal point for range O&M. This squadron also performs QA evaluation of range contractors with oversight from a senior QA evaluator

---

[1] Current PTR metrics include, but are not limited to, operational availability, system downtime, MTTR, maintenance frequency, range and equipment utilization rates, contractor discrepancy reports, and costs per system.

at ACC. Additional flights in the range squadron perform engineering and communications and computer services for the range mission.

Range equipment modernization and capability-enhancement initiatives are managed and prioritized through the combat training range's executive review board and realistic training review board, held semiannually. Besides A3A, the other key players at HQ ACC are: Training (A3T), for weapons issues and aircrew training requirements; Environmental (A7V), for range environmental impacts; and Communications (A6O), for equipment frequency issues. ACC does standardize range equipment across the PTRs when possible. Short-term operational requirements are addressed at both the CTR and RTRB. Long-term (typically three to five years) requirements are addressed in comprehensive range plans produced by the range operating agencies (wings).

For local initiatives, the range wings have local procurement authority up to the level of their annual financial plans.

For use of the PTRs by other DoD components and agencies, the Overarching Range Cooperative Agreement provides training services without fee, except when the requested range time would require overtime labor or when unique requirements exist. An example of the latter is the U.S. Navy's request to drop high-explosive weapons at a range the U.S. Air Force uses only for inert weapons.

Table E.1
Comparison of ACC and AFSPC Range Support Contracts

	ACC Primary Training Ranges	AFSPC Eastern and Western Ranges
Contract management	Centralized at the MAJCOM. Joint effort between ACC/A3A and ACC Contracting Squadron	Decentralized to SW RMSs
Scope	O&M	O&M
Contract type	Firm fixed price	Cost reimbursement
Metrics	System performance and contractor performance	System performance and contractor performance
Range management organization	Range squadron under the OG	RMS under the OG
QAE function	Decentralized with ACC oversight	Decentralized
Cost recovery for non-U.S. Air Force customers	Unique requirements only	Unique requirements only
Modernization initiative review process	Semiannual combat training ranges executive review board and realistic training review board; held at MAJCOM	Periodic requirements validation boards are held at MAJCOM for initiatives exceeding the $500,000 threshold

Reliability-Centered Maintenance Prioritization Process

This appendix presents an example of a range system contractor's RCM prioritization process. Call Henry, Inc., prime contractor for the Western Range Launch Operations Support Contract (see Appendix D), leads the fleet in application of RCM to equipment and facility maintenance.

Call Henry identifies and prioritizes its maintenance workload using RCM techniques. The following example considers the decision on when to repair and/or replace the protective coating on range facilities. Figure E.1 shows the typical costs of replacing the mechanical coating (e.g., paint) on a facility across the spectrum of wear and failure modes.

This approach is then applied to specific facilities in the Western Range complex, using the five condition criteria in Figure F.2, providing an assessment of risk (likelihood of failure versus consequences and/or mission impact) and a weighted score used in prioritizing requirements against limited repair or replacement funding resources.

For each facility, Call Henry surveys and records condition factors, then assigns a composite rating based on severity from "acceptable" or "optimum" to "past due" or "critical." Figure F.3 is an actual summary for facilities at a launch complex.

Call Henry and the government jointly review and reprioritize the maintenance workload (as necessary) to achieve best use of limited funding in executing the wing's mission. In this way, RCM techniques provide a method of repairing and/or replacing facility coatings at the proper time and in the proper priority sequence.

Figure F.1
Mechanical Coating System Costs

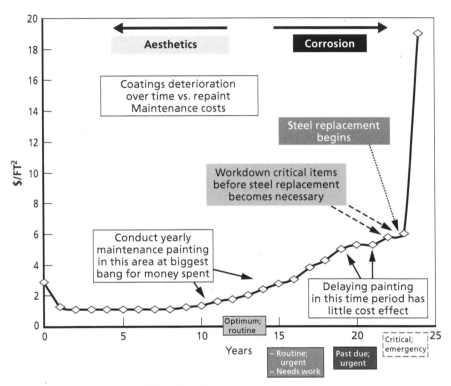

SOURCE: Call Henry, Inc. (2005, slide 31).
RAND *MG518-F.1*

Figure F.2
Condition Criteria and Risk Assessment

Looking at five condition criteria
- Environment—Type-Severity (Corrosion Rate)
- Visibility—Where Aesthetics Count
- Corrosion—Rust Grade, Wear, Adhesion
- Priority—Need & Cost Effectiveness, Lead
- Risk Level—Of Loss, Hazard, Mission Critical

Condition Analysis
"Weighting values"
- Risk 24.0%
- Priority 19.0%
- Visibility 19.0%
- Corrosion 19.0%
- Environment 19%

Applying the five condition assessment criteria

Environment

Rate	Description	Weight
Vs	Very severe	12
S	Severe	10
C	Coastal	8
M	Ext. Moderate–mild	5
I	Internal–slight	3

Visibility

Rate	Description	Weight
Vhv	High public visibility	12
Hv	Shop–work area	10
Mv	Moderate traffic or visibility	6
Lv	Low traffic or visibility	3
Vlv	No visibility	1

Priority

Rate	Description	Weight
Critical	Advanced corrosion	12
Optimum	Optimum time	12
Needs	Substandard coatings	8
Past due	Localized corrosion	7
Ok	Wear	3
Blast	No increased coats	3
Good	No work needed	1
Caretaker	Excluded	0

Corrosion–rust grade

Code	Description	Weight
1.0	>50% rust	12
2.0	33–50% rust	12
3.0	16–33% rust	10
4.0	10–16% rust	10
5.0	3–10% rust	8
6.0	1–3% rust	6
7.0	0.3–1% rust	4
8.0	0.1–0.3% rust	3
9.0	<0.1% rust	1

Risk consequence

Rate	Description	Weight
10	Severe (5 x 5)	15
9	Very high (5 x 4)	13
8	High (4 x 4 or 5 x 3)	10
6–7	Medium	6
5	Low	3
≤4	Slight	1

Risk Assessment Grid

Probability: 5 Very High, 4 High, 3 Moderate, 2 Low, 1 Very Low

Consequence: 1 Very low, 2 Low, 3 Moderate, 4 High, 5 Very high

SOURCE: Call Henry, Inc. (2005, slide 36).

RAND MG518-F.2

Figure F.3
Corrosion Survey Summary

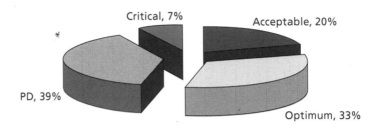

RAND *MG518-F.3*

21st Maintenance Group

Considering the various organizational structures of space wings in 14AF, this appendix briefly describes the single example that implements a maintenance group structure.[1] The 21 SW, which provides space surveillance, is, at least at the group level, organized in accordance with the CWO into operations, maintenance, mission support, and medical groups.

Although system-specific maintenance management is "embedded" in each operations squadron (similar to 50 OG SOPS), several maintenance functions have been consolidated into 21 MXG to provide a partial strategies-to-task implementation. Also of note, the PMD organization performs tasks similar to the range management squadrons in 30 OG and 45 OG. The maintenance group comprises the units shown in Table G.1

---

[1] All three SWs in 20AF, the numbered air force under AFSPC that operates and maintains the nation's ICBM force, are task-organized into MXGs. Each MXG consists of a Missile Maintenance Squadron and Maintenance Operations Squadron consolidating maintenance functions into a single group.

Table G.1
21st Maintenance Group Subordinate Units

Unit	Function	Authorized Manpower				
		Officers	Enlisted	Civilian	Contractor[a]	Total
Command Section and Staff	Overall direction and leadership for the MXG	3	6	2	0	11
21 SCS	Operates and maintains communications and computer systems for wing and higher headquarters missions	14	113	85	97	309
21 MOF	Provides maintenance site support, depot-level maintenance coordination, system status tracking, maintenance data analysis, technical data support, and evaluation for precision measurement equipment laboratories, transient aircraft maintenance, and aerospace ground equipment services	2	20	2	23	47
PMD	Manages 10 major programs (including requirements, reacquisition, cost, performance, schedule, and award fees) and provides QA support, training, and management	1	0	30	0	31
Total		20	139	119	120	398

[a] Although contractor positions are not "authorized" per se, these data provide a measure of staff augmentation and the total requirement.

Evolution of Space Wing Maintenance

In the space wings of 14AF, maintenance management functions have undergone several organizational changes since the early 1990s. This appendix describes a few of the changes and the reasons for realignment.

In the early 1990s, space-system maintenance functions were divided between the LG and OG. Organizational-level maintenance was embedded in operational squadrons (e.g., the space operations squadron, space warning squadron, and SLS) under the operations group, creating strong partnerships with operations, with the squadron commander serving as integrator at the unit level. Cross-cutting maintenance functions (e.g., intermediate-level and groupwide activities) along with supply, transportation, and contracting, made up the LG. This aligned closely with the objective wing structure created by former Chief of Staff Gen Merrill A. McPeak.

Following the Chief's Logistics Review, primarily an aircraft-centric study, HQ U.S. Air Force issued Program Action Directive 02-05, directing the change to the CWO. Under the CWO, by October 1, 2002 the maintenance resources embedded in operational squadrons were to be realigned into new aircraft maintenance squadrons under the maintenance group, the successor to the LG. The logistics functions of supply, transportation, and contracting were realigned to the mission support group. The maintenance group consisted of two to three maintenance squadrons and a maintenance operations squadron, with all aircraft maintenance centralized under a single O-6 commander, the maintenance group commander.

In the fall of 2002, AFSPC's space wings in 14AF reorganized in response to Program Action Directive 02-05, divesting the LG's logistics functions and the contracting squadron to the mission support group. At the range wings, the new maintenance group was left with only two squadrons: an SCS and a new range management squadron. However, unlike flying wings, the range wings did not break apart their SLSs, which retained their embedded maintenance. At 21 and 50 SWs, the operations squadrons similarly retained their embedded maintenance. Maintenance is embedded in the system program office for 460 SW. In addition, while the space wing formed a maintenance group, the 50 and 460 SWs did not.

In December 2003, after prompting by the Space Commission and the broad-area review, AFSPC again reorganized its space wings in 14AF. This reorganization, aimed at tightening the integration between operations and acquisition and reducing or eliminating seams for placing assets on orbit, either intentionally or unintentionally did away with the maintenance group structure at all but one wing. One senior leader indicated that, after the CWO was implemented, the maintenance group (former LG) was considerably smaller, making it a candidate for realignment—which occurred. At the range wings, the SCS and range management squadron units in the fledgling maintenance groups were realigned into the operations group, and SLSs, still with their embedded maintenance, were realigned out of the operations group into the newly formed launch control groups. At 50 SW, the operations group remained largely unchanged, but the network operations group was created to consolidate communications maintenance management. For 460 SW, no significant changes took place. The 21 MXG emerged the "lone maintenance group standing" in its present structure (see Appendix G). The wing commander at the time, a former Director of Logistics and Communications at HQ AFSPC, successfully retained the continued existence of the maintenance group.

Since that time, several minor adjustments have been made at the various space wings. The launch groups have divested their "heritage" launch vehicle systems, deactivating two SLSs and creating two new launch support squadrons. The 50 SW is planning to merge its two SCSs.

Bibliography

14th Air Force Staff, Summary Sheet, "RAND Space Command Maintenance Review: Balancing Current and Future Capabilities," March 2006.

Air Force Magazine: 2005 Air Force Almanac, special ed., Vol. 88, No. 5, May 2005. As of December 6, 2006:
http://www.afa.org/magazine/May2005/

Assistant Deputy Under Secretary of Defense for Maintenance Policy, Programs, and Resources, "CBM+ Plan of Action and Milestones," rev., Washington, D.C.: U.S. Department of Defense, August 2004.

Call Henry, Inc., "Project Management Review (PMR)," briefing, August 29, 2005.

Cebrowski, Arthur, and John Raymond, "Operationally Responsive Space: A New Defense Business Model," *Parameters*, Summer 2005.

Commission to Assess United States National Security Space Management and Organization, *Report of the Commission to Assess United States National Security Space Management and Organization*, January 11, 2001.

Dahlman, Carl, and David Thaler, *Assessing Unit Readiness: Case Study of an Air Force Fighter Wing*, Santa Monica, Calif.: RAND Corporation, DB-296-AF, 2000. As of September 19, 2006:
http://www.rand.org/pubs/documented_briefings/DB296/

Defense Acquisition University, *Defense Acquisition Guidebook*, Fort Belvoir, Va., May 12, 2003. As of August 31, 2006:
http://akss.dau.mil/dag/DoD5000.asp?view=document

Drew, John G., Russell Shaver, Kristin F. Lynch, Mahyar Amouzegar, and Don Snyder, *Unmanned Aerial Vehicle End-to-End Support Considerations*, Santa Monica, Calif.: RAND Corporation, MG-350-AF, 2005. As of September 19, 2006:
http://www.rand.org/pubs/monographs/MG350/

Gabreski, Terry L., *Chief of Staff, United States Air Force Logistics Review*, Washington, D.C.: Headquarters, AF/ILM, June 2000.

Jumper, John P., "Posturing Aircraft Maintenance for Combat Readiness," briefing to the Chief of Staff of the Air Force, September 1999.

Kent, Glenn, *A Framework for Defense Planning*, Santa Monica, Calif.: RAND Corporation, R-3721-AF/OSD, 1989. As of September 19, 2006:
http://www.rand.org/pubs/reports/R3721/

Kiziah, Rex, *Joint Warfighting Space Vision and Technologies*, Air Force Research Laboratory, June 6, 2005.

Lewis, Leslie, James A. Coggin, and C. Robert Roll, Jr., *The United States Special Operations Command Resource Management Process: An Application of the Strategy-to-Tasks Framework*, Santa Monica, Calif.: RAND Corporation, MR-445-A/SOCOM, 1994. As of September 19, 2006:
http://www.rand.org/pubs/monograph_reports/MR445/

Lynch, Kristin F., John G. Drew, David George, Robert S. Tripp, Charles Robert Roll, Jr., and James A. Leftwich, *The Air Force Chief of Staff Logistics Review: Improving Wing-Level Logistics*, Santa Monica, Calif.: RAND Corporation, MG-190-AF, 2005. As of September 19, 2006:
http://www.rand.org/pubs/monographs/MG190/

NASA Office of Logic Design, *Space Launch Vehicles Broad Area Review Report*, briefing, November 5, 1999. As of September 19, 2006:
http://klabs.org/richcontent/Reports/Failure_Reports/Space_Launch_Vehicles_Broad_Area_Review.pdf

Niblack, Preston, Thomas S. Szayna, and John Bordeaux, *Increasing the Availability and Effectiveness of Non-United States Forces for Peace Operations*, Santa Monica, Calif.: RAND Corporation, MR-701-OSD, 1996.

Schrader, John Y., Leslie Lewis, William Schwabe, C. Robert Roll, and Ralph Suarez, *USFK Strategy-to-Task Resource Management: A Framework for Resource Decisionmaking*, Santa Monica, Calif.: RAND Corporation, MR-654-USFK, 1996. As of June 5, 2007:
http://www.rand.org/pubs/monograph_reports/MR654/

Snyder, Don, and Patrick H. Mills, *Sustaining Air Force Space Systems: A Model for the Global Positioning System*, Santa Monica, Calif.: RAND Corporation, MG-525-AF, 2007. As of June 5, 2007:
http://www.rand.org/pubs/monographs/MG525/

Space Commission—*See* Commission to Assess United States National Security Space Management and Organization.

Staats, Raymond W., and Derek A. Abeyta, "Technical Education for Air Force Space Professionals," *Air & Space Power Journal*, Vol. XIX, No. 4, Winter 2005. As of September 19, 2006:
http://www.airpower.maxwell.af.mil/airchronicles/back.htm

Thaler, David E., *Strategies to Tasks: The Framework for Linking Means and Ends*, Santa Monica, Calif.: RAND Corporation, MR-300-AF, 1993. As of September 19, 2006:
http://www.rand.org/pubs/monograph_reports/MR300/

Tripp, Robert S., Kristin F. Lynch, Charles Robert Roll, Jr., John G. Drew, and Patrick Mills, *A Framework for Enhancing Airlift Planning and Execution Capabilities Within the Joint Expeditionary Movement System,* Santa Monica, Calif.: RAND Corporation, MG-377-AF, 2006. As of September 19, 2006:
http://www.rand.org/pubs/monographs/MG377/

Walker, James V., "30 RMS/RMQ Quality Assurance Flight," briefing to RAND, September 2005.

Official Documents

Air Force Doctrine Documents
2-2, *Space Operations*, November 17, 2001.

2-4, *Combat Support*, March 23, 2005.

Air Force Instructions
10-602, *Determining Mission Capability and Supportability Requirements*, March 18, 2005.

21-103, *Equipment Inventory, Status, and Utilization Reporting*, February 25, 2005 (more recent edition: December 14, 2005).

21-108, *Maintenance Management of Space Systems*, July 25, 1994.

21-116, *Maintenance Management of Communications-Electronics*, March 24, 2006.

21-203, *Space Maintenance Management*, draft, undated.

38-101, *Air Force Organization*, April 4, 2006.

Air Force Policy Directive
10-12, *Space*, February 1, 2006.

Other
Air Force Space Command Instruction 10-1208, *Launch and Range Roles and Responsibilities*, November 1, 2005.

Technical Order 00-20-2, "Maintenance Data Documentation," June 15, 2003.

"Guidelines for Reliability, Maintainability, and Availability (RMA) Metrics for the Launch Test Range System (LTRS)," rev., October 2005.

Headquarters, U.S. Air Force, "Implementation of the Chief of Staff of the Air Force Direction to Establish a New Combat Wing Organization Structure," Program Action Directive 02-05, June 20, 2002.